Linear Algebra
and Its
Applications

SERIES ON UNIVERSITY MATHEMATICS

ISSN: 1793-1193

Editors:

Wu-Yi Hsiang	*University of California, Berkeley, USA/*
	Hong Kong University of Science and Technology,
	Hong Kong
Tzuong-Tsieng Moh	*Purdue University, USA*
Ming-Chang Kang	*National Taiwan University, Taiwan (ROC)*
S S Ding	*Peking University, China*
M Miyanishi	*University of Osaka, Japan*

Published

SERIES ON UNIVERSITY MATHEMATICS – VOL. 10

Linear Algebra
and Its
Applications

Tzuong-Tsieng Moh

Purdue University, USA

World Scientific

NEW JERSEY · LONDON · SINGAPORE · BEIJING · SHANGHAI · HONG KONG · TAIPEI · CHENNAI · TOKYO

Published by

World Scientific Publishing Co. Pte. Ltd.
5 Toh Tuck Link, Singapore 596224
USA office: 27 Warren Street, Suite 401-402, Hackensack, NJ 07601
UK office: 57 Shelton Street, Covent Garden, London WC2H 9HE

Library of Congress Cataloging-in-Publication Data
Names: Moh, T. T., author.
Title: Linear algebra and its applications / by Tzuong-Tsieng Moh (Purdue University, USA).
Description: New Jersey : World Scientific, 2019. | Series: Series on university mathematics :
    volume 10 | Includes index.
Identifiers: LCCN 2019002091 | ISBN 9789813235427 (hardcover : alk. paper)
Subjects: LCSH: Algebras, Linear--Problems, exercises, etc.
Classification: LCC QA184.5 .M64 2019 | DDC 512/.5--dc23
LC record available at https://lccn.loc.gov/2019002091

British Library Cataloguing-in-Publication Data
A catalogue record for this book is available from the British Library.

For any available supplementary material, please visit
https://www.worldscientific.com/worldscibooks/10.1142/10861#t=suppl

Typeset by Stallion Press
Email: enquiries@stallionpress.com

Printed in Singapore

TO

My wife Ping

Preface

Linear Algebra is the most important subject in Mathematics after *Calculus*. It has a long history starting with the ancient Chinese mathematics book *Nine Chapters on Arithmetic*. In there, the solutions of a system of linear equations were given. Later on, in 1683 AD, Japanese mathematician Seki Kōwa[1] defined the concept of the *determinant*. After the 18th century, the center of research on linear algebra was moved to Europe. Then the algebraic method was studied in Europe, and there was an explosion of knowledge. In the 17th century, Newton and in the 18th century, Gauss started to study the solving of systems of linear equations. Gauss recovered the knowledge of the *Nine Chapters on Arithmetic*, and the methods bear the name of Gauss as the Gaussian elimination methods, etc. The usage of linear algebra spreads out to cover all fields of sciences. In this book on *linear algebra*, we will start with discussing the solving of a system of linear equations, the algebraic structures of matrices, the theoretical aspects of the theory of matrices, the applications of matrices to *mathematics, physics, engineering, and business*, and the numerical computation methods of matrices. We also venture into the area of infinite dimensional vector spaces. In there, the theory of matrices is ineffective. We have to use the finite dimensional knowledge as a guiding light only. Our main tool there is *set theory*, especially *Zorn's lemma*.

---

[1]Kōwa, Seki. Japanese mathematician. 1642–1708.

The present book is based on a compilation of lecture notes from a course for first-year graduate students at Purdue University. We present the details of the book as follows. The lecture notes have six chapters: Chapter 1 is about solving a system of linear equations. In Section 1.1, we present historical facts; this may be traced back to the ancient Chinese Mathematical book, *Nine Chapters on Arithmetic* (10th century BC to 2nd century BC), which gave a method to find the reduced row echelon form to solve a system of linear equations (the method is now called the *Gaussian row operations*). Naturally we introduce *vector spaces*, especially *finite dimensional vector spaces* and matrix theory. We prove the natural and important theorem of the *Normal Form* of matrices, and the existence of basis for a general vector space. We apply this section to the original *self-correcting codes*, the *Hamming codes*. In Section 1.2, we discuss *linear transformations*, especially we discuss natural examples of linear transformations. In Section 1.3, we discuss the geometric meaning of the classical dot product. In this way, we will introduce *geometry* to the real or complex vector spaces. We will prove the classical *Gram–Schmidt theorem* about the existence of orthonormal basis. In Section 1.4, we discuss the *fundamental theorem* of matrices in the sense of G. Strang and its application to the electric network in the form of *Kirchhoff's law*. In Section 1.5, we review the elementary definition of *determinants*. We give a preliminary discussion about the m-dimensional volume in $\mathbb{R}^n$ where $m \leq n$. In Section 1.6, we use *set theory* to establish the dimension theory of general vector spaces.

In the 18th century, several European mathematicians developed the *algebraic theory* of matrices. In Chapter 2, we will repeat their journey. In Section 2.1, we will extend the *abstract ideas*, which generalize the numerical sets from fields K to rings R, and in Section 2.2, we extend the objective sets from vector spaces V to modules M. In Section 2.3, we discuss a vector space-like object *free module* and discuss an interesting example of *double periodic function* in *complex analysis*. In Section 2.4, we have *finitely generated modules over a P.I.D.* and define the *Smith Normal Form* of a matrix. In Section 2.5, we show that the Smith Normal Form is of interest by applying it to *coding theory*. In Section 2.6, we have *finitely generated modules over a P.I.D.*, and we study the fundamental theorem of it. In Section 2.7, we are given a square matrix A, we define the *rational form* and the *characteristic polynomial* of it. In Section 2.8, we will show that given a square matrix A, we use the Smith Normal Form to deduce the *torsion decomposition* and *elementary decomposition* for A. Moreover, the decompositions are unique for a given A. In Section 2.9, we show the existence of the *Jördan Canonical Form* in two different ways. We apply it to solve the following system of differential equations,

$$X' = AX$$

where $X = [x_1, \ldots, x_n]^T$ is a vertical vector function of unknowns $\{x_i\}$. Note that this is an important part of designing a modern *microchip*. In Section 2.10, we discuss *eigenvalues, eigenvectors* and prove the classical *Cayley–Hamilton theorem*. We apply the concepts and prove the classical *Euler rotation theorem* for $\mathbb{R}^{2n+1}$. In Section 2.11, we discuss the phenomena of *simultaneously diagonalizing* finitely many square matrices. We show a deep connection of it to the algebraic property of commutativity of multiplication. In the future Sections 2.5 and 2.6, we will connect it to the *Uncertainty Principle* in Quantum Mechanics.

In Chapter **3**, we discuss determinants. In Section 3.1, we present the modern *axiomatic definition* of the determinant and the classical *Laplace formula*. From the modern definition, we deduce a fast way of computation of the determinant and the identification of the determinant with the n-dimensional volume in an n-dimensional vector space $\mathbb{R}^n$. In Section 3.2, we show a way of computing the m-dimensional volume in an n-dimensional space $\mathbb{R}^n$ where $m \leq n$. We take this as an introduction to the classical *Cauchy–Binet theorem*. In Section 3.3, we present the classical and modern definitions of *tensor product* and the *exterior product*. We show that the exterior product defines all m-dimensional volume vectors. We apply the tensor product to differential geometry by making computations on a torus in $\mathbb{R}^3$. In Section 3.4, we introduce the fundamental *dual space* of V, and discuss the relation between the double dual and the space V. We copy an example from Feynman's book to illustrate the usage of tensor product and dual space in physics.

In Chapter **4**, we study inner product spaces. In Section 4.1, we introduce *inner product spaces*, and show that length and angle are the consequence of *inner product*. As usual, we give examples and counterexamples. We define by integration an inner product structure of function spaces. In Section 4.2, we show the *Gram–Schmidt theorem* for any dimensional inner product space. We discuss the *least square approximation, trigonometric series* and *Fast Fourier Transform* (FFT) as examples of *perpendicular projection*. In Section 4.3, we study some elementary theory of *Hilbert space*. In Section 4.4, we study the *perpendicular complementary subspace* of a closed subspace of a Hilbert space. In Section 4.5, we study *adjoint* and *self-adjoint operators*. After von Neumann, we show the subtle difference between self-adjoint and *hermitian operators*. We apply this material of inner product space to geometry and get classification theories of the quadratic curves in $\mathbb{R}^2$, and quadratic surfaces in $\mathbb{R}^3$. We apply the results to *Quantum Mechanics* to deduce the famous *Uncertainty Principle*. In Section 4.6, we study the extreme problems in *Calculus* of many variables, and find the formula of the second derivative tests for three or more variables. Thus we complete a chapter on *Calculus*. In Section 4.7, we discuss the *unitary operator*. In Section 4.8, we discuss the *spectrum theorem*.

In Chapter **5**, we study *bilinear forms* and the decomposition. In Section 5.1, we discuss various bilinear forms. We define *bilinear form, symmetric bilinear form*, etc. We establish *Sylvester's law of inertia*, the invariant of *signatures*. In Section 5.2, we study the group which keeps a given bilinear form invariant, and apply those concepts to physics and geometry. We study the *special relativity* and the *Lorentz Group SO(1,3)*. In Section 5.3, given a rectangular matrix A, we discuss the *singular value decomposition* (SVD) of it. We discuss a meaningful *low rank approximation* which can be considered as a way to eliminate *noises*, and can be applied to the film processing and *latent semantic analysis* in information retrieving. Note that there exist several softwares on the market to compute SVD.

In Chapter **6**, we study numerical computations of matrices. In Section 6.1, we study numerical models, the finite difference method and the finite element method. In Section 6.2, we study non-negative square matrices, and prove the *Perron–Frobenius Theorem*, and show its application to the *Leontief Economic Model* and the *Google search engine* using the power method. We show that it is meaningful and possible to compute

$$I + A + A^2 + \cdots = (I - A)^{-1}$$

and

$$\lim_{n \mapsto \infty} A^n \mathbf{1}$$

where $\mathbf{1}$ is the vector with all components 1, and show its application to the Google search engine. In Section 6.3, we study the QR method to approximate the *upper-triangular decomposition* of a square matrix (Schur's lemma). We introduce the reader to the *Hessenberg matrix* and *Householder transformation* and their applications to computational linear algebra. In Section 6.4, we study some fast numerical approximation to the solutions of a system of equations of sizes $> 10^6$ variables. Especially, we present the well-known *Conjugate Gradient Method*.

In the last section of this book, we study the well-known subject of *linear programming*. We discuss the usual *simplex method*, the duality theorem and a description of the interior point methods.

We hope that this book will help students to appreciate the beautiful subject of *linear algebra*. There are finite dimension vector spaces, which are foundations of the theory of *matrices* and *determinants*. There are infinite dimension vector spaces, which are illustrated by *inner product spaces* and *Hilbert spaces* mainly. Furthermore, the subject can be classified as *theory, application* and *numerical computation*. We write on all lines of developments. We weave the several lines into a chord. And, we hope the readers will enjoy it.

We wish to express our appreciation to Prof. Mark Ward for drawing four pictures in this book for us.

We wish to thank Ms. Rochelle Kronzek, executive editor at World Scientific Publishing Company (WSPC), for her constant enthusiasm on initializing this project. We are grateful to Ms. Lai Fun Kwong, managing editor at WSPC, for her prompt communications and support during the book writing. I am indebted to Mr. Rajesh Babu at WSPC, for his tireless help during the production process. I also would like to thank Ms. Patricia Huesca Ignacio at Purdue University, for her numerous assistance in completing this manuscript.

Contents

Chapter 1

Preliminaries

In ancient times, humans studied the equation

$$ax = b.$$

The solution is $x = a^{-1}b$. There are two lines of development:
(1) one goes to a higher degree in one variable, (2) one goes to a
system of linear equations in many variables. Certainly this fuses into
modern *Algebraic Geometry* which studies the systems of equations
involving higher degree polynomials of many variables. The ancient
Greeks studied curves of higher degrees. And the ancient Chinese
mathematical book, *Nine Chapters on Arithmetic* (10th century BC–
2nd century BC), presented systems of linear equations in many
variables and used *row operations* to produce the reduced row echelon
form to solve a system of linear equations. This method is usually
called *Gaussian row operations* now. We adopt the usual name
for it.

In this chapter, we firstly discuss finite dimensional vector spaces
and review the undergraduate materials and especially the method of
solving systems of linear equations which is the topic in *elementary
linear algebra*. The most useful tool is the *Gaussian row operations*.
The important result is Theorem 1.1.6, that is about the *normal form*
of a matrix A. Secondly, we study some basic facts about arbitrary
dimensional vector spaces.

1.1   Echelon Forms and Matrix Algebras.
Hamming Codes

Echelon Forms

We assume that the readers are familiar with undergraduate linear algebra, the real numbers $\mathbb{R}$ and the complex numbers $\mathbb{C}$. The real space $\mathbb{R}^n$ for some positive integer n is the collection of all n-tuples $[r_1, r_2, \ldots, r_n]^T$, i.e.,

$$
\mathbb{R}^n = \left\{ \begin{bmatrix} r_1 \\ r_2 \\ \cdot \\ \cdot \\ r_n \end{bmatrix} : r_i \in \mathbb{R} \right\}.
$$

An $m \times n$ matrix A is as follows,

$$
A = \begin{bmatrix}
a_{11} & a_{12} & \cdots & \cdots & a_{1n} \\
a_{21} & a_{22} & \cdots & \cdots & a_{2n} \\
\cdots & \cdots & \cdots & \cdots & \cdots \\
\cdots & \cdots & \cdots & \cdots & \cdots \\
a_{m1} & a_{m2} & \cdots & \cdots & a_{mn}
\end{bmatrix}.
$$

In the above, the coefficients r_i, a_{ij} are real numbers at the beginning, and then they are generalized to complex numbers in the 19th century. Later on, the numbers are allowed to be general. Especially, after the invention of computers, sometimes the coefficients are from the set of two elements $\{0, 1\}$. We first start with a basic mathematical object and consider only one binary operation. We have the following definition of *Group*,

Definition (Group): Let G be a set. A binary operation $\cdot$ is a rule to assign an element $c = a \cdot b$, given any pair of elements (a, b) in G. If $c = a \cdot b \in G$ always, then we say that G is *closed* under the binary operation $\cdot$. If G is closed under $\cdot$ and has the following three properties, then we say G forms a *group* under the binary operation:

(1) (Associative law): $(a \cdot b) \cdot c = a \cdot (b \cdot c)$.
(2) The existence of identity: there exists an element e such that $a \cdot e = e \cdot a = a$.

(3) The existence of inverses: for any a, there exists an element b such that $a \cdot b = b \cdot a = e$.

Sometimes we omit the mention of $\cdot$, if it is obvious, and we simply say that G is a group. If $a \cdot b = b \cdot a$ always, we say that G is an *abelian group* or a *commutative group*. ∎

Now we consider a general concept of a *field* which is defined by Dedekind[1] as follows,

Definition (Field): A non-empty set K together with two operations $+, \times$ (we may say $(K, +, \times)$) form a *field* if:

(1) The set K is a *commutative group* with respect to $+$. We name the unit element of addition as 0.
(2) The set $K \backslash \{0\}$ is a non-empty commutative group with respect to $\times$. Let us denote the unit as 1. We further define $a \times 0 = 0 \times a = 0$ for all $a \in K$.
(3) The relations between the two operations $+$ and $\times$ are distributive, i.e., $k_1 \times (k_2 + k_3) = k_1 \times k_2 + k_1 \times k_3$ for all $k_i \in K$. ∎

A good definition calls for good examples and good counterexamples. We have the following

Example 1.1: The real numbers $\mathbb{R}$ and the complex numbers $\mathbb{C}$ are fields. In the study of computer science, we meet the *field* of two elements $\{0, 1\}$,

$$0 + 0 = 0, 1 + 0 = 0 + 1 = 1, 1 + 1 = 0$$
$$1 \times 1 = 1, 0 \times 0 = 1 \times 0 = 0 \times 1 = 0.$$

On the other hand, the set of integers $\mathbb{Z}$ is not a field, for instance, the number 2 has no multiple inverse as an integer. ∎

There are many usages of *field* theory as follows.

Example 1.2: We extend analytic geometry by replacing real numbers $\mathbb{R}$ as an axis to any field K. Let us consider the following

---

[1]Dedekind, R. German mathematician. 1831–1916.

system of equations with coefficients $a, b, c, d, e, f \in K$,

$$ax + by = c$$

$$dx + ey = f$$

with $\det\left|\begin{smallmatrix} a & b \\ d & e \end{smallmatrix}\right| = ae - bd \neq 0$.

Then the system of equations has a unique solution pair. Geometrically, we extend the usual *analytical geometry* to *algebraic geometry*. Instead of real numbers, we use any field as an axis. In this way, we think that each linear equation defines a *linear space*. The condition on the *determinant* means that the two lines are not parallel. The unique solution of the system of equations is the unique common intersection point. We thus fully generalize the real plane to a 2-dimensional plane with coordinate axis in a given field K. We denote the 2-dimensional plane by A_K^2 (with variables x, y) over any field K and call it the *affine plane over K*. The major difference between the affine plane A_K^2 and 2-dimensional vector space (for a definition, see below) is that in the affine plane there is no particular point named the origin 0 as uniquely defined for a vector space. In general, we have for each non-negative integer n, an n-dimensional affine space A_K^n. ∎

The preceding example shows that given a field K, we have an analytical geometry built on it. We may turn the table and show that given an irreducible geometric object, we have a natural field of rational functions associated with it. Let us consider the simplest cases.

Example 1.3: Let K be any field and A_K^n be the n-dimensional affine space over K. The field of rational functions of n-variables is the set $K(x_1, \ldots, x_n) = \left\{ \frac{f(x_1, \ldots, x_n)}{g(x_1, \ldots, x_n)} : g(x_1, \ldots, x_n) \neq 0 \right\}$ and is called a *rational function field* of n variables. If we assign the variables $x_1, \ldots, x_n$ to be a coordinate system of A_K^n, then we shall call $K(x_1, \ldots, x_n)$ the rational function field of A_K^n. ∎

The just-mentioned exchanges of geometric objects and function fields are not idle games of ivory towers, for instance, a problem in calculus can be solved in the following example.

Example 1.4 (Rational Function Field): Let C be an *irreducible algebraic plane curve* defined by an irreducible polynomial equation $f(x, y) = 0$. The curve C is said to be *rational* if the equation $f(x, y) = 0$ can be in parametric form by the following non-constant rational functions

$$x = g(t)$$
$$y = h(t)$$

such that $f(g(t), h(t)) = 0$. For instance, say C is defined by $f(x, y) = x^2 + y^2 - 1$. The curve $C : x^2 + y^2 - 1 = 0$ can be in parametric form by t as a projection from the point $(0, 1)$,

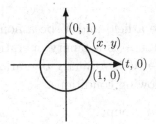

It is left to verify that

$$x = \frac{2t}{t^2 + 1}$$
$$y = \frac{t^2 - 1}{t^2 + 1}.$$

We may apply the above discussion to a problem of integration: find

$$\int_0^{2\pi} f(\sin\theta, \cos\theta)d\theta$$

where $f(x, y)$ is a rational function of two variables. Note that we shall set $x = \sin\theta, y = \cos\theta$, then $x^2 + y^2 = 1$ and

$$d\theta = \frac{dx}{y}$$
$$dx = \frac{2 - 2t^2}{(t^2 + 1)^2}dt.$$

Therefore the problem of trigonometric integration can be transformed into an integration of rational functions as

$$\int f(\sin\theta, \cos\theta)d\theta = \int g(t)dt$$

where $g(t)$ is a rational function of t.   ∎

Descartes and Fermat were the first to describe vectors in $\mathbb{R}^2$ and $\mathbb{R}^3$. Later on physicists used vectors to describe forces, velocities etc. Then analysts studied the solution sets of linear differential equation systems which are vector spaces. In 1888, G. Peano[2] called his vector spaces *linear systems*. We have the following common definition of *vector space*.

Definition Let $\mathbf{K}$ be a field and $\mathbf{V}$ be a non-empty set. A *vector space* $(\mathbf{V}, +, \cdot, \mathbf{K})$, where $+$ is a binary operation between elements in $\mathbf{V}$ and $\cdot : \mathbf{K} \times \mathbf{V} \mapsto \mathbf{V}$ is a binary operation with $a, b \in \mathbf{K}$ and $v \in \mathbf{V}$ satisfies the following conditions,

(1) $(V, +)$ is an abelian group;
(2) *Associative law:* $(a \cdot (b \cdot v)) = (a \cdot b) \cdot v$, for all a, b, v;
(3) *Distributive law:* $(a+b) \cdot v = a \cdot v + b \cdot v$ and $a \cdot (v+u) = a \cdot v + a \cdot u$;
(4) $1 \cdot v = v$;

where the elements in V are called *vectors*, and the field K is called the *ground field*. Sometimes we omit the mention of $\cdot$, if it is obvious.   ∎

It is very useful to have a concept of *dimension* for a space. For vector spaces, we use the *cardinality* (see below) of a *basis* as the dimension of a vector space. There are different dimensions defined for different branches of mathematics. For instance, we have *covering dimension* or *Lebesgue dimension* in *analysis*, and *Hausdorff dimension* for *fractal geometry* and the *Krull dimension* in *polynomial ring* by a chain of prime ideals.

If we have two fields $\mathbf{L} \supset \mathbf{K}$, then clearly $\mathbf{L}$ is a vector space over $\mathbf{K}$.

---

[2]Peano, G. Italian mathematician and glottologist. 1858–1932.

We have the following definitions.

Definition Let $S = \{v_i\}_{i \in I}$ be a subset of a vector space V. (1) If every element $v \in V$ can be written as a finite linear combination of elements in S, i.e.,

$$v = \sum_{\text{finite}} a_i v_i$$

where $a_i \in K$ the ground field, then S is said to be a *generating set*. (2) If elements in S only satisfies the trivial relation, i.e.,

$$0 = \sum_{\text{finite}} a_i v_i \iff a_i = 0 \; \forall i,$$

then S is said to be a *linearly independent set*. If it is not a linearly independent set, then it is a *linearly dependent set*. (3) The set S is both a generating set and a linearly independent set, iff S is a *basis*. ∎

Definition A vector space V is said to be *finite dimensional* iff there is a finite basis in V. ∎

The concept of basis follows from the concept of coordinate system of Descartes.[3] For any coordinate system on a vector space V with the origin of the coordinate system being the zero vector of the vector space, we take v_i as any non-zero vector on the ith axis. Then we have the definition of a basis for a vector space. For a coordinate system, it must have the existence and uniqueness properties. Namely, every vector v can be expressed as $v = \sum_{\text{finite}} a_i v_i$, then the vector v has a coordinate $\{a_i\}$. Furthermore the expression must be unique. Namely if $v = \sum_{\text{finite}} a_i v_i = \sum_{\text{finite}} b_i v_i$, then we must have $a_i = b_i$ for all i. This is the linearly independent property (see the Exercises).

We have the following proposition for the existence of a basis.

Proposition 1.1.1: *A maximal linearly independent set is a basis. A minimal generating set is a basis.*

---

[3]Descartes, R. French philosopher, mathematician and physicist. 1586–1650.

Proof. The terms maximal and minimal are terms in the *set theory.* *A set S is a maximal linearly independent set* means that it is a linearly independent set and if thrown in any elements to S, then it will lose the property of being a linearly independent set. *The set S is a minimal generating set* means that it is a generating set and if we throw out any elements from S, then it will lose the property of being a generating set.

Let us prove the first half of the proposition. Let S be a maximal linearly independent set. Let v be any element in V. By the maximal property, we must have a non-trivial relation,

$$av + \sum_{finite} a_i v_i = 0.$$

If $a = 0$, then it becomes a relation among linearly independent set S. Therefore the relation must be trivial. If $a \neq 0$, then the relation becomes

$$v = \sum_{finite} (-a_i/a)v_i.$$

Hence S is a generating set, therefore S is a basis.

The second part is left to the reader as an exercise. ∎

Naturally given any vector space V, we want to show the existence of a basis. To establish it, we have to use the basic *set theory* which is discussed below.

In the usual axiom system of set theory, we have the *axiom of choice*: which states that a non-empty product of non-empty sets is non-empty. For instance, let us consider $\prod_{i \in I} S_i = \{f : f(i) \in S_i\}$ where I is a non-empty index set. If S_i's are pairs of shoes, then clearly $\prod_{i \in I} S_i$ is not empty, because we have $f(i) =$ the left shoe, hence the product is not empty. On the other hand, if S_i's are pairs of socks, then we need the axiom of choice to say the product is not empty. The axiom of choice may be replaced by the following *Zorn's*[4] *lemma* in the usual system of axioms of set,

---

[4]Zorn, M. German–American mathematician. 1906–1993.

Lemma 1.1.2 (Zorn's Lemma): *Let F be a non-empty family of partial ordered sets. If every linear ordered subfamily has an upper bound in F, then there is a maximal element in F.* ∎

The lemma itself needs no proof because it is equivalent to the *axiom of choice*. In the statement we use the term *family* which is synonymous with the term *set*. We have to explain the other terms used above. We have,

Definition A binary relation $\leq$ is said to be a *partial order* if $a \leq b$, $b \leq c$, then $a \leq c$. ∎

For instance, all subsets of a set form a partially ordered family by inclusion.

Definition A partial order $\leq$ is said to be a *linear order* if for any two elements a, b, we must have $a \leq b$ or $b \leq a$. ∎

For instance, all subsets of a set usually is not a linear ordered set by inclusion and all integers form a linear order set under the usual inequality.

Definition Let S be a partial ordered set with a subset T. An element b is said to be an *upper bound* of T, iff $a \leq b$ for all $a \in T$. ∎

For instance, in the integers $\mathbb{Z}$, any integer $a \geq -1$ is an upper bound for all negative integers, while the subset of all positive integers is without upper bound.

Definition Let S be a partial ordered set. A *maximal element a* is an element such that $a \leq b$ for any $b \in S$ implies $a = b$. ∎

Now we have the following proposition which proves that there is a basis in any vector space V.

Proposition 1.1.3: *Given any vector space $V \neq \{0\}$, and a linearly independent set A, then there is a basis $S \supset A$.*

Proof. We shall define a family F and a partial ordered $\leq$. Let $F = \{S_i : S_i$ linearly independent set, and $S \subset S_i\}$. It is easy to see $F \neq \emptyset$. Let us define $S_i \leq S_j$ iff $S_i \subset S_j$. Then it is easy to see that $\leq$ is a partial ordering.

Let us verify the hypothesis of Zorn's lemma. Let $\{S_i\}$ be a linear order subfamily. Let $S = \cup_i S_i$. We want to show that S is a linearly independent set. Suppose not. Then there is a non-trivial relation

$$\sum_j a_j v_j = 0.$$

It is easy to see that there is a k such that $v_j \in V_k$ for all j. Therefore the above equation is over elements in V_k. Contradiction! Henceforth we show that a maximal element is in F, and it is a linearly independent set. Our proposition follows from the preceding Proposition 1.1.1. ∎

Let us start our journey cautiously. Let us start with a special kind of finite dimensional vector space.

Notations: Given a field K, we use the notation $\mathbf{K}^n$ to denote the following vector space,

$$K^n = \left\{ \begin{bmatrix} k_1 \\ k_2 \\ \cdot \\ \cdot \\ k_m \end{bmatrix} : k_j \in K \right\}$$

where the summation of two vectors is defined to be,

$$\begin{bmatrix} k_1 \\ k_2 \\ \cdot \\ \cdot \\ k_m \end{bmatrix} + \begin{bmatrix} k_1' \\ k_2' \\ \cdot \\ \cdot \\ k_m' \end{bmatrix} = \begin{bmatrix} k_1 + k_1' \\ k_2 + k_2' \\ \cdot \\ \cdot \\ k_m + k_m' \end{bmatrix}$$

and the scalar multiplication $r \cdot v$ is defined to be

$$r \cdot v = r \cdot \begin{bmatrix} k_1 \\ k_2 \\ \cdot \\ \cdot \\ k_m \end{bmatrix} = \begin{bmatrix} r \cdot k_1 \\ r \cdot k_2 \\ \cdot \\ \cdot \\ r \cdot k_m \end{bmatrix}.$$

We use the notation $\mathbf{K}_n$ to denote the following vector space in a similar way,

$$K_n = \left\{ [k_1, k_2, \ldots, k_n] : k_j \in K \right\}. \quad \blacksquare$$

We want to generalize the theorems of *linear algebra* allowing any *field* as numerals. Later on when we discuss the concepts of *inner product*, *length* and *angle* for vectors, we have to use the real numbers $\mathbb{R}$, or the complex numbers $\mathbb{C}$ only. Otherwise we will use any field for the coefficients in our discussions.

We have the following definition for (Gaussian[5]) *reduced echelon form* for matrices.

Definition Given a matrix A as

$$A = \left\{ \begin{bmatrix} R_1 \\ R_2 \\ \cdot \\ \cdot \\ R_m \end{bmatrix} : R_j \text{ row vector of length } n \right\}.$$

The matrix A is said to be in a *reduced row echelon form* (1) If all R_j's are of the form $[0, \ldots, 0, 1, \ldots]$ where the first non-zero term is 1 and happens at n_jth position. Usually this non-zero term is called the *pivot term*. (2) If $i < j$ then $n_i < n_j$. (3) If on the n_jth column, only that particular coefficient is 1, all other coefficients are zeroes. $\quad \blacksquare$

Similarly, we define the *reduced column echelon form* as follows.

Definition Given a matrix A as

$$A = [C_1, C_2, \ldots, C_n] : C_j \text{ column vector of length } m.$$

---

[5]Gauss, C. German mathematician and physicist. 1777–1855.

The matrix A is said to be in a *reduced column echelon form* (1) If all C_j's are of the form $[0, \ldots, 0, 1, \ldots]^T$ where the first non-zero term is 1 and happens at m_jth position. Usually this non-zero term is called the *pivot term*. (2) If $i < j$ then $m_i < m_j$. (3) If on the m_jth row, only that particular coefficient is 1, all other coefficients are zeroes. ∎

Let $A = [R_1, \ldots, R_m]^T$, $B = [R_1', \ldots, R_m']^T$ be two matrices. If one of the following conditions is satisfied: (1) There are $i \neq j$, such that $R_i = R_j'$, $R_j = R_i'$ and $R_k = R_k'$, $\forall k \neq i, j$. (2) There is i and a non-zero $a \in K$ such that $R_i = aR_i'$ and $R_k = R_k'$, $\forall k \neq i$. (3) There are $i \neq j$ such that $R_j' = R_j + bR_i$ for some $b \in K$, and $R_k' = R_k$, $\forall k \neq j$, then we say B is a row transform of A. It is easy to see that the subspaces spanned by $\{R_1, \ldots, R_m\}$ and $\{R_1', \ldots, R_m'\}$ are equal.

The main theorem for the reduced row echelon forms in the elementary linear algebra is as follows,

Theorem 1.1.4: *Using a sequence of elementary row operations, we may transform any matrix B to a reduced row echelon form.*

Proof. Elementary linear algebra. ∎

Similarly we have the dual theorem,

Theorem 1.1.5: *Using a sequence of elementary column operations, we may transform any matrix B to a reduced column echelon form.* ∎

The following theorem claims the uniqueness of the reduced row echelon form,

Theorem 1.1.6: *Given a matrix B. If by different sequences of elementary row operations we have two reduced row echelon forms A and A', then $A = A'$.*

Proof. Let $A = [R_1, \ldots, R_m]^T, A' = [R_1', \ldots, R_m']^T$ where R_i, R_i' are row vectors. We want to show that $R_i = R_i'$. First there are numbers $n_1, \ldots, n_m, n_1', \ldots, n_m'$ where n_i is the position of the leading non-zero term of R_i (similarly n_i' is the position of the leading non-zero term for R_i') in the definition of the reduced echelon form. We claim $n_i = n_i'$ for all i.

Let us start with $i = 1$. If $n_1 \neq n_1'$. We may assume that $n_1 < n_1'$. Clearly each of the sets $\{R_1, \ldots, R_m\}, \{R_1', \ldots, R_m'\}$ spans the row space of B. Hence we have the following equation,

$$R_1 = a_1 R_1' + \cdots + a_n R_n'.$$

Look at the coefficient of the position of n_1. Then at the n_1's position the left-hand side is a vector with coefficient 1, while every component $a_i R_i'$ has coefficient 0. We get a contradiction. Therefore we conclude that $n_1 = n_1'$. Inductively we may assume that $n_j = n_j'$ for all $j < i$. Let us assume further that $n_i < n_i'$ and $n_i \leq n$ (otherwise the vectors R_i' are the 0 vector). As before we have the following equation

$$R_i = a_1 R_1' + \cdots + a_n R_n'.$$

If one of the coefficients $a_1, \ldots, a_{i-1}$ is non-zero, then we get a contradiction by comparing the leading non-zero term on both sides of the equation. Hence we may assume $a_1 = a_2 = \cdots = a_{i-1} = 0$. Then we have the following equation,

$$R_i = \sum_{j \geq i} a_j R_j'.$$

It is easy to deduce that $n_i = n_i'$ and $a_i = 1$. By mathematical induction, we conclude that $n_i = n_i' \ \forall i$. We have

$$R_i - R_i' = \sum_{j > i} a_j R_j'.$$

If the right-hand side is not 0, let k be the minimal index such that $a_j \neq 0$. Then the coefficient at n_k position is 0 on the left-hand side, while non-zero at the right-hand side. A contradiction. Therefore the right-hand side of the above equation is zero, and $R_i = R_i'$. ∎

The proof of the following theorem is the dual of the proof of the preceding theorem.

Theorem 1.1.7: *Given a matrix B. If by different sequences of elementary column operations we have two reduced column echelon forms A and A', then $A = A'$.* ∎

We have the following useful proposition.

Proposition 1.1.8: *Given any square $n \times n$ matrix A. Then there is a product of elementary matrices E_i such that $(\prod_i E_i)A$ is either identity matrix I or the last row vector is the zero vector.*

Proof. See Exercises.   ∎

We have the following proposition which is a dual of the preceding proposition,

Proposition 1.1.9: *Given any square $n \times n$ matrix A. Then there is a product of elementary matrices E_i such that $A(\prod_i E_i)$ is either identity matrix I or the last column vector is the zero vector.* ∎

We have the following proposition of solving equations,

Proposition 1.1.10: *Given a system of equations $Ax = b$. Let E be the product of elementary matrices such that EA is the reduced echelon form B of A. Let $Bx = EAx = Eb = c$ be an equivalent system of equations. Let*

$$
B = \begin{bmatrix} R_1 \\ R_2 \\ \cdot \\ R_{m-1} \\ R_m \end{bmatrix}, \qquad c = \begin{bmatrix} c_1 \\ c_2 \\ \cdot \\ c_{m-1} \\ c_m \end{bmatrix}.
$$

Let k be the number such that $R_j \neq 0$ for $j \leq k$ and $R_j = 0$ for all $j > k$. Then the system has a solution iff $c_j = 0$ for all $j > k$.

Proof. Elementary linear algebra.   ∎

We may further consider the column operations which amount to multiplying elementary matrices from the right.

Normal Form

A matrix $M_{m \times n}$ is said to be in *normal form* if

$$
M = \begin{bmatrix} I_{k \times k} & 0_{k \times n-k} \\ 0_{m-k \times k} & 0_{m-k \times n-k} \end{bmatrix}
$$

where $I_{k \times k}$ is a $k \times k$ identity matrix, and $0_{i \times j}$ is a $i \times j$ zero matrix.

We have the following theorem,

Theorem 1.1.11 (Normal Form): *We shall apply the multiplications of the elementary matrices on the reduced echelon form of A from the right, then after finitely many steps, the echelon form CA of A is further transformed to CAD which is in the normal form of A.*

Proof. See the previous discussions. ∎

One of the main theorems of the next chapter is the *Smith Normal Form* of a matrix over a P.I.D. Note that a field is a P.I.D., and the *Smith Normal Form* is a generalization of the *normal form* over a field.

Matrix Algebra

The subject *algebra* is the study of arithmetic operations on a set. Every matrix $A_{m \times n}$ has a *type* $m \times n$. In ancient times, it was known that two matrices A, B can be added only if their types are the same. The multiplications of matrices were developed in Europe in the 18th century. After that, *linear algebra* started flying high. We multiply only matrices $A_{\ell \times m}, B_{m \times n}$. Recall the following definitions,

Definition Let $A_{\ell \times m} = (a_{ij}), B_{m \times n} = (b_{jk})$ be two matrices. Then we define the product $AB = C = (c_{ik})$ where

$$c_{ik} = \sum_j a_{ij} b_{jk}.$$ ∎

Definition Let A, B be square matrices. If $AB = I$ is the identity element, then A is said to be a left inverse of B, and B is said to be a right inverse of A. If there are matrices A, C such that $AB = I$ and $BC = I$, then B is said to be invertible. ∎

We have the following propositions,

Proposition 1.1.12: *If a square matrix B is invertible, i.e., there are matrices A, C such that AB = I = BC, then A = C. We will call A an inverse of B. Moreover the inverse of B is unique.*

Proof. We have $C = (AB)C = A(BC) = A$. ∎

Proposition 1.1.13: *All $n \times n$ square invertible matrices form a group.*

Proof. Trivial. ∎

Proposition 1.1.14: *If a square matrix B has a left inverse, then the equation $Bx = b$ can be solved for any vector b.*

Proof. Let the left inverse be A. Multiply the equation by A from the left, we get

$$x = ABx = Ab = c.$$

Therefore we solved for the variable vector x. ∎

We have the following proposition,

Proposition 1.1.15: *For any square matrix A, if it has a left inverse then it is invertible.*

Proof. Suppose that a square matrix $B_{n \times n}$ is the left inverse of $A_{n \times n}$. We shall find a matrix E which is a product of elementary matrices such that EA is either the identity or the last row 0 vector. We will show that it is impossible to have the last row 0 vector.

Let E be the product from left of all elementary matrices which correspond to the Gaussian row operations which reduce A to its reduced row echelon form. Since E is the product of invertible matrices corresponding to Gaussian row operation, it is invertible. Let its inverse be E^{-1}. Let

$$c = \begin{bmatrix} 0 \\ 0 \\ \cdot \\ \cdot \\ 1 \end{bmatrix}$$

and $E^{-1}c = b$. Then the equation $Ax = b$ is equivalent to $EAx = Eb = c$. The last row of the system of equations is of the form

$$0 = 1.$$

Hence it cannot be solved and it contradicts to a preceding proposition. We conclude that $EA = I$. Hence $A = E^{-1}$ which is invertible.

∎

Proposition 1.1.16 (Gauss–Jordan[6]): *Let A be an $n \times n$ square matrix. Let us form an $n \times 2n$ matrix $[A|I]$. Let $\overline{A} = (\prod_i E_i)A$ be the reduced row echelon form of A. Then $\prod_i E_i I$ is the inverse of A iff $\overline{A} = I$.*

Proof. We know that either $\overline{A} = I$ or the last row is the 0 vector. In the second case, the inverse of A does not exist, therefore $\prod_i E_i I$ cannot be the inverse of A. In the first case, the inverse of A is clearly $\prod_i E_i = \prod_i E_i I$.

∎

Let us use the following notation:

Notation: Let V be a finite dimensional vector space over a field K. Let $E = \{e_1, \ldots, e_m\}$ be a subset of V. We write $\mathbf{e} = [e_1, \ldots, e_m]^T$. Let $N = \{n_1, \ldots, n_m\}$ be a finite set of V. We write $\mathbf{n} = [n_1, n_2, \ldots, n_n]^T$. Let $n_j = \sum_{i=1}^{m} a_{ij} e_j$. We write $[n_1, \ldots, n_n]^T$ as $(a_{ij})\mathbf{e}$ where (a_{ij}) is an $m \times n$ matrix A. We write $\mathbf{n} = A\mathbf{e}$.

∎

Lemma 1.1.17: *Let C, D be invertible matrices. (1) If E is a generating set for a subspace U of V, then $C\mathbf{e}$ is a generating set for U. (2) If E is a linearly independent set, then $C\mathbf{e}$ is a linearly independent set. (3) If E is a basis for V, then $C\mathbf{e}$ is a basis for V.*

Proof. In the following, let $\mathbf{c} = [c_1, \ldots, c_m]$ be a horizontal vector in K^m. (1) For any vector v, we have $v = \sum c_i e_i = \mathbf{c}\mathbf{e} = (\mathbf{c}C^{-1})(C\mathbf{e})$ where $\mathbf{c}$ is a horizontal vector in K^m. (2) If $\mathbf{c}(C\mathbf{e}) = 0$, then $\mathbf{c}(C\mathbf{e}) = (\mathbf{c}C)\mathbf{e} = 0$. Therefore $\mathbf{c}C = 0$. Since C is invertible, then $\mathbf{c} = 0$. (3) follows from (1) and (2).

∎

We will state some consequence of the *normal form* of a matrix.

---

[6]Jordan, W. German geodesist. 1742–1899.

Proposition 1.1.18: (1) *Let V be a vector space with a finite generating set* **e**. *Then the cardinality of the set* **e** *is greater or equal to the cardinality of any linearly independent set* **n**. (2) *Suppose that a vector space V has a finite generating set. Then any basis of V has the same finite cardinality.*

Proof. Let the relation matrix between **n** and **e** be A, i.e., $\mathbf{n} = A\mathbf{e}$. Let the invertible matrices C, D be given such that $DAC = N$ is in the normal form of A. Then we have $D\mathbf{n} = DACC^{-1}\mathbf{e} = DAC\mathbf{e}' = N\mathbf{e}'$, where $\mathbf{e}' = C^{-1}\mathbf{e}$ is another generating set, and $D\mathbf{n}$ is another linearly independent set, hence it contains no zero vector. Since the normal form N is an $m \times n$ matrix and the last row is not zero, we must have $n \geq m$.

(2) follows from (1) by considering the first basis as a generating set, and the second basis as a linearly independent set and then switch the roles.   ■

The above proof of the invariant of the cardinality of any basis in a finite dimensional vector space can be generalized to infinite dimensional cases (see Section 1.6).

We apply our knowledge of $\mathbb{R}^n$ to construct *secret sharing* as follows.

Example 1.5 (Secret Sharing): Suppose we want to design a way for each individual in a group to have some information and the whole information can be found only if sufficiently many of them work together.

We may apply our knowledge of $\mathbb{R}^n$ to achieve this purpose. For instance, let the secret be a point in R^2, and suppose each individual has a distinct line passing through it. With only one line, the point cannot be found, while if two of them get together, then they can solve the system of linear equations to find the point. In general, we may consider R^n. Let the secret be a point in R^n. We give each individual in a group of m ($m \geq n$) people a co-dimension 1 hyperplane in general positions. Then only if n or more people get together can they determine the point.

Suppose that there is a secret safe with the code [123456]. The code may be given to two people with the understanding that the code is given by a point (x, y) where x is the first part of the digits and y is the second part of the digits of the code. Each of the two gets an envelope containing one of the two equations,

$$y - x = 333$$
$$y - 2x = 210.$$

Each person with one equation cannot find the code, while two people collaborate to locate the secret. ∎

We state some consequences of the *normal form* of a matrix.

Proposition 1.1.19: (1) *Let V be a vector space with a finite generating set $\mathbf{e}$. Then the cardinality of the set $\mathbf{e}$ is greater or equal to the cardinality of any linearly independent set $\mathbf{n}$. (2) Suppose that the vector space V has a finite generating set, then it has a finite minimal generating set with respect to inclusion. (3) Suppose that a vector space V has a finite basis. Then any basis of V has the same finite cardinality.*

Proof. (1)(a) Let the relation matrix between $\mathbf{n}$ and $\mathbf{e}$ be A, i.e., $\mathbf{n} = A\mathbf{e}$. Let the invertible matrices C, D be given such that $DAC = N$ is in the normal form of A. Then we have $D\mathbf{n} = DACC^{-1}\mathbf{e} = DAC\mathbf{e}'$, where $\mathbf{e}' = C^{-1}\mathbf{e}$ is another generating set, and $D\mathbf{n}$ is another linearly independent set, hence $D\mathbf{n}$ contains no zero vector. Since the normal form N is an $m \times n$ matrix and the last row is not zero, we must have $n \geq m$.

(b) We use the *replacement principle*. The importance of this method is that it works for *infinite cases*. Let the generating set be $\{v_i : i \in I\}$, and the linearly independent set be $\{u_j : j \in J\}$. We have the following equation

$$u_1 = \sum_{finite} a_i v_i$$

with at least one of $a_i \neq 0$, we assume that $a_1 \neq 0$. Then we **claim** that $\{u_1\} \cup \{v_i : i \neq 1\}$ is a generating set.

Proof of the Claim. Let $v \in V$ and $v = \sum_{finite} b_i v_i$. If $b_1 = 0$, then we are clearly done. If $b_1 \neq 0$, then we have

$$v = b_1 v_1 + \cdots = b_1((-1/a_1)u_1 + \cdots).$$

Therefore our **claim** is proved.

In a general step of mathematical induction, suppose that $\{u_j : j \notin S\} \cup \{u_j : j \in T\}$ is a generating set where $S \subset I, T \subset J$ are two index subsets with the same cardinalities. We have used elements with indices in T to replace elements with indices in S. We want to carry the process one step further.

If $J \backslash T = \emptyset$, (i.e., $J = T$), then we have $Card(I) \geq Card(S) = Card(T) = Card(J)$. We are done. If $J \backslash T \neq \emptyset$, then let $k \in J \backslash T$. Let

$$u_k = \sum_{j \notin S} a_j v_j + \sum_{j \in T} b_j u_j.$$

It is impossible to have all $a_j = 0$, since then it will be a non-trivial equation among elements in J which is a set of linearly independent elements. There must be an element $a_i \neq 0$. We use u_k to replace v_i. We prove that after replacement, we still have a generating set. The proof is the same as the previous proof.

(2) If the finite generating set S is minimal, then we are done. Otherwise there is a subset which is a generating set. The statement is obvious by a mathematical induction of the cardinality of the set S.

(3) Follows from (1) by considering the first basis as a generating set, and the second basis as a linearly independent set and then switch the roles. ∎

The above proposition can be generalized to the following,

Proposition 1.1.20: *Let V be a vector space. Then the cardinality of a generating set* **e** *is greater or equal to the cardinality of any linearly independent set* **n**. *Any two bases of V will have the same cardinality.*

Proof. A proof of the above proposition involves *set theory* which is a distraction of our present task and will be presented in *Section 1.5* for the interested readers. ∎

Example 1.6: In 1945, a computer scientist, R. W. Hamming, used a primitive (by today's standard) computer to perform his research. At that time, one had to queue one's work for the computer to a sequential process. If the computer found errors, say typos, in the program, it would skip the task and precede to the next one in the queue. The researcher would have to correct the errors in the faulty program, resubmit it to the queue, and wait several weeks for the computer to find time to work on it again. Apparently, Hamming was annoyed by the waiting and decided to create a "self-correcting code" which eventually bore his name, as follows:

Let us consider $GF(2)$ $(= \mathbf{F}_2)$, the field of two elements $\{0, 1\}$ which will be called the set of *alphabet*, and message $a_1 a_2 a_3 a_4$ where $a_i \in \{0, 1\}$. Hamming added three more symbols $b_1 b_2 b_3$ by the following formula,

$$
\begin{aligned}
b_1 &= a_1 + a_3 + a_4 \\
b_2 &= a_1 + a_2 + a_3 \\
b_3 &= a_2 + a_3 + a_4.
\end{aligned} \tag{1}
$$

Then he used seven symbols $a_1 a_2 a_3 a_4 b_1 b_2 b_3$ to carry the message of four symbols $a_1 a_2 a_3 a_4$. We may consider the following matrix multiplication, with

$$
G = \begin{pmatrix}
1 & 0 & 0 & 0 & 1 & 1 & 0 \\
0 & 1 & 0 & 0 & 0 & 1 & 1 \\
0 & 0 & 1 & 0 & 1 & 1 & 1 \\
0 & 0 & 0 & 1 & 1 & 0 & 1
\end{pmatrix}
$$

and $[a_1 a_2 a_3 a_4] \times G = [a_1 a_2 a_3 a_4 b_1 b_2 b_3]$. The matrix G is called the *generator matrix*. The a_i's are called the *message symbols*. Furthermore, let

$$
H = \begin{pmatrix}
1 & 1 & 0 \\
0 & 1 & 1 \\
1 & 1 & 1 \\
1 & 0 & 1 \\
1 & 0 & 0 \\
0 & 1 & 0 \\
0 & 0 & 1
\end{pmatrix}
$$

and $[a_1 a_2 a_3 a_4] \times G \times H = [000]$. The matrix H is called the *check matrix* and b_i's are called the *check symbols*.

The decoding process is as follows. Suppose that the computer, for whatever reasons, reads $[a_1 a_2 a_3 a_4 b_1 b_2 b_3]$ as $[a_1' a_2' a_3' a_4' b_1' b_2' b_3']$ which might be different from the original string. However this kind of error is infrequent, so we may reasonably assume that there is at most one error, i.e., either

$$[a_1' a_2' a_3' a_4' b_1' b_2' b_3'] = [a_1 a_2 a_3 a_4 b_1 b_2 b_3]$$

or

$$[a_1' a_2' a_3' a_4' b_1' b_2' b_3'] = [a_1 a_2 a_3 a_4 b_1 b_2 b_3] + [0 \cdots 010 \cdots 0].$$

The computer calculates

$$[a_1' a_2' a_3' a_4' b_1' b_2' b_3'] \times H = [c_1 c_2 c_3].$$

If $[c_1 c_2 c_3] = [000]$, then the above defining equations (1) for b_1, b_2, b_3 show that there is either no error or there are more than two errors. Since we assumed that there is at most one error, we may conclude there is no error, so the computer should take the message $[a_1' a_2' a_3' a_4']$. If $[c_1 c_2 c_3] \neq [000]$, then we have

$$\begin{aligned}
[a_1' a_2' a_3' a_4' b_1' b_2' b_3'] &\times H \\
&= ([a_1 a_2 a_3 a_4 b_1 b_2 b_3] + [0 \cdots 010 \cdots 0]) \times H \\
&= [0 \cdots 010 \cdots 0] \times H \\
&= [c_1 c_2 c_3].
\end{aligned}$$

Therefore $[c_1 c_2 c_3]$ must be one of the row vectors of the matrix H, and thus the computer locates the position of the error. The computer simply flips the bit at that position. In this way the computer will correct the code $[a_1' a_2' a_3' a_4' b_1' b_2' b_3']$ and takes the first four bits of the corrected message as the message. This code not only detects the error but also corrects the error. Note that the *location* of the error must be detected before it can be corrected, a process that will be referred to as *error-locator*.

However, there might be two or more errors, in which case the Hamming code fails and the above method decodes the message to the wrong word. One may assume that the possibility is rather small. The usage of Hamming codes is to eradicate a single error, but it is ineffective if the errors are multiple.

Let us consider the successful rate of Hamming code. Let us assume that $p =$ the rate for symbols $0, 1$ to be transmitted incorrectly, and $q = 1 - p$ is the rate the symbols are transmitted correctly. Then we have

$$(p + q)^7 = q^7 + C_1^7 q^6 p + \cdots .$$

Therefore we conclude that the Hamming code is effective for a probability $q^7 + C_1^7 q^6 p$.

For a longer message, we can chop it into blocks (each with four bits), padding the end if necessary. This produces a *block* code.

The principles of self-correcting in Hamming codes are valid today, and widely used in communications through noisy channels. Academically, a **code** means a **self-correcting code**. All channels of communications are noisy to different degrees. The self-correcting codes have become prominent today. ∎

We have the following proposition,

Proposition 1.1.21: *Given a system of equations $Ax = b$. We use a sequence of elementary matrices E_i to multiply A to its reduced row echelon form B (i.e., let $E = \prod_i E_i$, then $EA = B$). The system of equations $Bx = EAx = Eb = c$ is equivalent to the original system $Ax = b$. Let k be defined as the integer k such that in the expressions,*

$$B = \begin{bmatrix} R_1 \\ R_2 \\ \cdot \\ R_{m-1} \\ R_m \end{bmatrix}, \quad c = \begin{bmatrix} c_1 \\ c_2 \\ \cdot \\ c_{m-1} \\ c_m \end{bmatrix}.$$

$R_j \neq 0$ for all $j \leq k$, and $R_j = 0$ for all $j > k$. If $c_j = 0$ for all $j > k$, then there are solutions.

Proof. Elementary linear algebra. ∎

Exercises

(1) Solve the following linear system of equations:
$$x + 2y + 3z = 6$$
$$2x + 3y + 2z = 5$$
$$3x + y + z = 5.$$

(2) Show that any $n + 1$ vectors $v_1, v_2, \ldots, v_{n+1}$ in $\mathbb{R}^n$ must be linearly dependent.

(3) Show that reduced row echelon form of an $n \times n$ matrix A is either I or with the last row 0 vector.

(4) Are the following two matrices row equivalent?
$$A = \begin{bmatrix} 1 & 2 \\ 2 & 3 \end{bmatrix}, \qquad B = \begin{bmatrix} 1 & 2 \\ 2 & 2 \end{bmatrix}.$$

(5) Use the Gauss–Jordan method to find the inverse of A, where
$$A = \begin{bmatrix} 1 & 2 & 0 \\ 2 & 3 & 2 \\ 1 & 3 & 3 \end{bmatrix}.$$

(6) Show that a subset $\{v_i\}$ is a linearly independent set iff $\sum_{finite} a_i v_i = \sum_{finite} b_i v_i$, then $a_i = b_i$ for all i.

(7) Show that if $\{v_i\}$ is a basis, then $\{v_i\}$ is a maximal linear independent set.

(8) Let V be a vector space with U, W its subspaces. Suppose that $V = U \cup W$. Show that either $V = U$ or $V = W$.

(9) Let U be the subspace of $\mathbb{C}^{2n}$ with the first n entries 0. Let W be the subspace of $\mathbb{C}^{2n}$ with $\xi_j = \xi_{j+n}$ where $j \leq n$ and ξ_k is the ith coordinate of the element in $\mathbb{C}^{2n}$. Show that $\mathbb{C}^{2n} = U \oplus W$.

(10) Find the conditions for scalars a, b, c such that $[1, a, a^2]^T$, $[1, b, b^2]^T, [1, c, c^2]^T$ are linearly dependent.

(11) Show that a minimal generating set is a basis for a vector space.

(12) Prove Proposition 1.1.8.

(13) Find the reduced row echelon form of A where $\mathbb{A}$ is the following matrix,
$$A = \begin{bmatrix} 1 & 2 & 3 \\ 2 & 3 & 2 \\ 2 & 4 & 1 \end{bmatrix}$$

(14) Show that $\mathbb{Q}$ is not finitely generated over $\mathbb{Z}$. (Hint: Use Zorn's lemma.)

(15) Find the normal form of A as

$$A = \begin{bmatrix} 1 & 2 & 3 \\ -1 & -3 & 4 \\ 2 & 1 & 1 \end{bmatrix}.$$

(16) Let V be a finite dimensional vector space, and ϕ an operator of V. Show that ϕ is surjective $\Longleftrightarrow \phi$ is injective.

(17) (Cayley's Parametric Representation) Show that every complex number z with $|z| = 1$ and $z \neq -1$, can be written uniquely as

$$z = (1 + ri)(1 - ri)^{-1}$$

where r is a real number.

(18) Write the inverse of the following matrix A as a product of elementary matrices, and use it to find the inverse of A,

$$A = \begin{bmatrix} 1 & 2 \\ 2 & 3 \end{bmatrix}.$$

(19) Let L be a line passing through $(1, 2, 3)$ with direction $(1, 1, 1)$. Let P be a plane passing through $(0, 0, 0)$ with two vectors $(1, 1, 0)$ and $(0, 1, 1)$ on it. Find the intersection of the line and the plane.

(20) Find the reduced echelon form of the following matrix

$$A = \begin{bmatrix} 1 & 2 & 3 & 4 \\ 5 & 6 & 7 & 8 \\ 9 & 10 & 11 & 12 \end{bmatrix}.$$

(21) Find the matrix which maps $\mathbb{R}^3$ to $\mathbb{R}^2$ and sends $[1, 2, 3]^T$ to $[1, 2]^T$, $[2, 3, 4]^T$ to $[2, 3]^T$ and $[3, 4, 5]^T$ to $[3, 4]^T$.

(22) Two matrices A, B are said to be row equivalent iff there is a sequence of row operations which changes A to B. Show that two matrices A, B are row equivalent iff their reduced row echelon forms are identical.

(23) Let the field be the real field $\mathbb{R}$. Show the following matrices are not row equivalent,

$$A = \begin{bmatrix} 1 & 2 & 3 & 4 \\ 5 & 6 & 7 & 8 \\ 9 & 10 & 11 & 12 \end{bmatrix}$$

$$B = \begin{bmatrix} 1 & 2 & 3 & 4 \\ 5 & 11 & 7 & 8 \\ 9 & 10 & 11 & 12 \end{bmatrix}.$$

(24) Let F be a field of two elements and V a 4-dimensional vector space over F. How many different basis are there?

(25) Suppose that you are communicating with Dr Hamming using the check matrix of this section and you receive a message [1110111]. What is the correct message?

(26) Show that every string $[a_1' a_2' a_3' a_4' b_1' b_2' b_3']$ can be corrected by switching one digit.

(27) Construct a Hamming code for $\mathbf{F_2}^4$. (Hint: We construct strings $[a_1 a_2 \cdots a_{11} b_1 b_2 b_3 b_4]$.)

1.2   Linear Transformations

One of the basic concepts of *Calculus* is the concept of *functions* which are maps from $\mathbb{R}$ to $\mathbb{R}$. We shall generalize it to the concept of *linear transformations* which is defined as follows,

Definition   Let V, U be two vector spaces over the same field K. A map $\rho : V \mapsto U$ is said to be a *linear transformation* iff

- $\rho(av) = a\rho(v)$, for $a \in K$, $v \in V$.
- $\rho(v_1 + v_2) = \rho(v_1) + \rho(v_2)$, for $v_1, v_2 \in V$.   ∎

In *Calculus*, we study maps from $\mathbb{R}$ to $\mathbb{R}$. Now we study maps from V to U, while we restrict maps to be linear only. Note that both V, U could be just K. Let us define the derivative of polynomials $K[x]$ as

Definition Given a polynomial $f(x) \in K[x]$, we define the derivative $f'(x)$ of $f(x) = \sum_{i \geq 0} a_i x^i$ as $f'(x) = \sum i a_i x^{i-1}$. ∎

Proposition 1.2.1: *The derivative is a linear transformation from $K[x]$ to $K[x]$.*

Proof. See any *Calculus* book. ∎

Let V be an n-dimensional vector space over a field K. Let $\mathbf{e} = \{e_1, \ldots, e_n\}$ be an ordered basis of V. Then $\mathbf{e}$ defines a map ρ from $V \mapsto K^n$ as follows, if $v = \sum a_i e_i$, then $\rho(v) = [a_1, a_2, \ldots, a_n]^T \in K^n$. We have the following proposition,

Proposition 1.2.2: *We use the notations of the preceding paragraph. Then ρ is a bijective (i.e., 1–1 and onto map) linear transformation.*

Proof. Exercise. ∎

The importance of the preceding proposition is that an abstract vector space V can be described numerically by a basis or a coordinate system, that is an important contribution of Descartes.[7] The algebraic method can be used to solve geometric problems.

In general, we have the *first isomorphism theorem* of Noether[8] of maps between vector spaces as follows,

Theorem 1.2.3: *Let $\rho: V \mapsto U$ be a linear map from vector space V to vector space U, and $N = \mathrm{Ker}(\rho), W = \mathrm{im}(\rho)$. Then it can be split into the following sequence*

$$0 \longrightarrow N \xrightarrow{\mu} V \xrightarrow{\pi} W \xrightarrow{\phi} U$$

---

[7]Descartes, R. French philosopher, mathematician and scientist. 1596–1650.
[8]Noether, E. Jewish–German mathematician and physicist. 1882–1935.

where μ, ϕ are embeddings, π is the canonical projection: $V \mapsto V/\mu(N)$, and $\rho = \pi\phi$. Hence especially

$$\dim(W) = \textit{the dimension of the image space} = \dim(V) - \dim(N).$$

Proof. The proof of the first part is identical with the proof for the same type of theorem about the abelian group. Let the kernel of the map ρ be N. Then it is easy to show that N is a vector space and a subspace of V. We have an embedding map of N to V. Then *image*(ρ) is canonically isomorphic to the quotient space $V/N = W$. Then the subspace W can be embedded to U.

Let us extend a basis $\{n_i\}$ of N to a basis $\{n_i\} \cup \{v_j\}$ of V. It is clear that

$$Card(\{v_j\}) = Card(\{n_i\}) + Card(\{v_j\}) - Card(\{n_i\})$$

hence our dimension formula. ∎

Let vector spaces V be n-dimensional, and U be m-dimensional. Let $v_1, \ldots, v_n$ be a basis for V, and $u_1, \ldots, u_m$ be a basis for U. Let ρ be a linear transformation from V to U. Certainly the linear transformation ρ is determined by the following data,

$$\rho(v_1) = \sum_j a_{1j} u_j$$

$$\rho(v_2) = \sum_j a_{2j} u_j$$

$$\ldots\ldots$$

$$\rho(v_n) = \sum_j a_{nj} u_j$$

since then given any $v = \sum_i b_i v_i$, we have $\rho(v) = \sum_i b_i \rho(v_i)$. We may write the above equations as $\rho(\mathbf{v}) = A\mathbf{u}$, where $\mathbf{v} = [v_1, \ldots, v_n]^T$ and $\mathbf{u} = [u_1, \ldots, u_m]^T$, the matrix $A = (a_{ij})$.

Proposition 1.2.4: *We have the following commutative diagram, i.e., $\beta\rho = A^T \alpha$,*

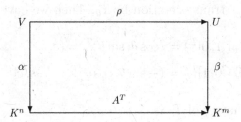

where α, β are the coordinate maps which send $v = \sum_i b_i v_i$, $u = \sum_j c_j u_j$ to $[b_1, \ldots, b_n]^T$, $[c_1, \ldots, c_m]^T$ respectively, i.e., $\alpha(v) = [b_1, \ldots, b_n]^T$, $\beta(u) = [c_1, \ldots, c_m]^T$.

Proof. Let $v = \sum_i b_i v_i$. Then we have

$$\beta\rho(v) = \beta\left(\sum_i b_i\rho(v_i) = \beta([b_1, \ldots, b_n]A\mathbf{u}) \right.$$

$$= \beta(([b_1, \ldots, b_n]A)\mathbf{u}) = A^T[a_1, \ldots, a_n]^T \left. \right) = A^T\alpha(v).$$

Therefore we conclude

$$\beta\rho = A^T\alpha. \qquad \blacksquare$$

Example 1.7: Let us consider simple cases of the above proposition. Let $V = U = \mathbb{R}^2$ and $v_1 = u_1 = [1, 0]^T, v_2 = u_2 = [0, 1]^T$. Any 2×2 matrix will define a linear transformation on $\mathbb{R}^2$. There are some interesting ones, the rigid motion (rotation and reflection) and projection. Let us discuss them separately.

Rotations: Let us rotate the whole plan by an angle θ around the origin as in the following picture,

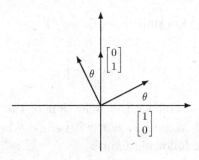

Let the linear transformation be T_θ. Then we have

$$T_\theta([1,0]^T) = (\cos\theta, \sin\theta)^T = (c,s)^T$$
$$T_\theta([0,1]^T) = (-\sin\theta, \cos\theta)^T = [-s,c]^T.$$

Therefore the matrix of T_θ is

$$A = \begin{bmatrix} c & -s \\ s & c \end{bmatrix}.$$

Projection: Let us project the whole plane to the line passing through the origin and with angle θ between it and the x-axis. We have the following figure:

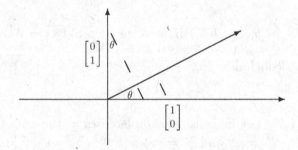

Let the linear transformation be P_θ. Then the image of any vector can be written as $r[\cos(\theta), \sin(\theta)]^T$, where r is the length of the image. We have

$$P_\theta([1,0]^T) = \cos(\theta)[\cos(\theta), \sin(\theta)]^T = [c^2, cs]^T$$

$$P_\theta([1,0]^T) = \sin(\theta)[\cos(\theta), \sin(\theta)]^T = [cs, s^2]^T$$

$$A = \begin{bmatrix} c^2 & cs \\ cs & s^2 \end{bmatrix}.$$

Reflection: Let us consider the reflection of the whole plan by a line passing through the origin and with angle θ between it and the x-axis. We have the following figure:

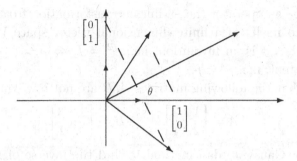

Let the linear transformation be H_θ. Then the image $H(v)$ of a vector v satisfies the following equation,

$$H(v) + v = 2P(v).$$

Therefore

$$H = 2P - I = \begin{bmatrix} 2c^2 - 1 & 2cs \\ 2cs & 2s^2 - 1 \end{bmatrix}.$$

Reflection through the Origin: Let us consider the reflection R of the whole plane through the origin. Every vector goes to its negative. We have the following figure:

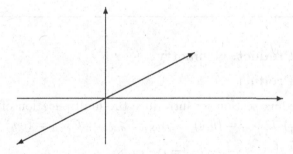

then the linear transformation R is $-I$. Therefore

$$R = -I = \begin{bmatrix} -1 & 0 \\ 0 & -1 \end{bmatrix}.$$

∎

Exercises

(1) Show that if V is an n-dimensional vector space, the map ρ in Proposition 1.2.2 of this section is bijective.

(2) Let ϕ be an operator (i.e., a linear transformation from a vector space to itself) of a finite dimensional vector space V over $\mathbb{R}$. Show that ϕ is an involution, i.e., $\phi^2 = I$, $\Leftrightarrow p = (1 + \phi)/2$ is idempotent, i.e., $p^2 = p$.

(3) Show that the following matrices A, B are not row equivalent,

$$A = \begin{bmatrix} 1 & 2 \\ 2 & 1 \end{bmatrix}, \qquad B = \begin{bmatrix} 1 & 3 \\ 3 & 1 \end{bmatrix}.$$

(4) Use the Gauss–Jordan method to find the inverse of matrix A over $\mathbb{R}$, where A is

$$A = \begin{bmatrix} 1 & 2 & 0 \\ 1 & 3 & 2 \\ 1 & 3 & 3 \end{bmatrix}.$$

(5) Find the matrix for $T_{\pi/6}$.

(6) Find the projection $P_{\pi/4}$.

(7) Find the reflection H for $\theta = \pi/3$.

(8) Show that $P_\theta^2 = P_\theta$.

(9) Let $C[0, 1]$ be all real continuous functions over the interval $[0, 1]$. Show that the multiple by x is a linear transformation from V to V.

1.3 Dot Product. Geometry

Real Dot Product

In the preceding section we introduce the multiplication of matrices $A_{\ell \times m} = (a_{ij}), B_{m \times n} = (b_{jk})$ as $A_{\ell \times m} B_{m \times n} = C_{\ell \times n} = (c_{ik})$ where

$$c_{\ell n} = \sum_j a_{ij} b_{jk}.$$

The product $\sum_j a_{ij} b_{jk}$ is meaningful in geometry.

Let us consider the real plane $\mathbb{R}^2$. The important geometric concepts are lengths and angles. Let $v = (a_1, a_2)^T$ be a vector, then its length $|v| = \sqrt{a_1^2 + a_2^2}$, this is the Pythagoras[9] theorem.

---

[9]Pythagoras of Samos. Greek philosopher, mathematician, well-known for metaphysics, music, mysticism, ethics, politics, and religion. 570–495 BC.

Similarly in $\mathbb{R}^n$, a vector $u = (b_1, b_2, \ldots, b_n)^T$ will have length $|u| = \sqrt{b_1^2 + b_2^2 + \cdots + b_n^2}$ (see Exercises). For angles θ, we take $\cos(\theta)$, this is similar to *Calculus*, where we take $\tan(\theta)$. What is the value of $\cos(\theta)$ where θ is the angle spanned by two vectors v, u? We have the generalized Pythagoras theorem or the cosine formula,

$$|c|^2 = |v|^2 + |u|^2 - 2|v||u|\cos(\theta)$$

where $|u|$ is the length of the vector $u = (a_1, a_2, \ldots, a_n)$ and $|v|$ is the length of the vector $v = (b_1, b_2, \ldots, b_n)$ and $|c|$ is the length of the third side $u - v$ of the triangle, and θ is the angle spanned by the two sides u, v. We have the following analytical equation,

$$\sum_i (a_i - b_i)^2 = \sum_i a_i^2 - 2\sum_i a_i b_i + \sum_i b_i^2 = |v|^2 + |u|^2 - 2|v||u|\cos\theta.$$

From the above equation, we deduce

$$\sum_i a_i b_i = |v||u|\cos\theta.$$

As usual we define $u \cdot v = u^T v = \sum_i a_i b_i$. Then the above equation can be simplified as

$$u \cdot u = |u|^2, \qquad u \cdot v = |u||v|\cos\theta.$$

Complex Dot Product

We may generalize the above to the complex dot product. Let us consider the complex space $\mathbb{C}^n$. Given two vectors $u = [c_1, c_2, \ldots, c_n]^T$, and $v = [f_1, f_2, \ldots, f_n]^T$, we define the dot product as $u \cdot v = \sum_j c_j \overline{f_j}$. Let $c_j = a_j + b_j i, f_j = d_j + e_j i$. Then we have

$$u \cdot v = \sum_j \det \begin{bmatrix} a_j & ib_j \\ ie_j & d_j \end{bmatrix} - i \sum_j \det \begin{bmatrix} a_j & b_j \\ d_j & e_j \end{bmatrix}.$$

It is not hard to see that $u \cdot u$ is a non-negative real number. The length is defined by *Pythagoras theorem* and can be written as

$$u \cdot u = |u|^2,$$

the angle θ can be computed by the *generalized Pythagoras theorem* as

$$|v - u|^2 = |v|^2 + |u|^2 - 2|v||u|\cos(\theta) = |v|^2 + |u|^2 - v \cdot u - u \cdot v;$$

we conclude that

$$\mathrm{Re}(v \cdot u) = |v||u|\cos(\theta)$$

and we further define $u \cdot v = 0 \implies u \perp v$, which means all vectors in $\mathbb{C}v$ are perpendicular to all vectors in $\mathbb{C}u$.

Remark: The vectors 1 and i in 1-dimensional space $\mathbb{C}$ span a $\pi/2$ angle while they are not perpendicular in our definition.

Gram[10]–Schmidt[11] Theorem

Definition A set $\{v_k\}_{k \in I}$ is said to be an orthogonal set if $v_k \neq 0$ for all i and $v_k \cdot v_j = 0$ for all $k \neq j$. An orthogonal set is said to be orthonormal set iff $|v_k| = 1, \forall k$. ∎

Proposition 1.3.1: *Let $u_1, u_2, \ldots, u_n$ be an orthonormal set and U be the subspace spanned by them. Then $u_1, u_2, \ldots, u_n$ forms a basis for U. For any vector $v \in V$, the projection of vector v to the subspace U is $P_U(v) = (v, u_1)u_1 + (v, u_2)u_2 + \cdots + (v, u_n)u_n$.*

Proof. It suffices to show that $w = v - \sum_k (v \cdot u_k)u_i$ is $\perp U$, which is obvious. Therefore $P_U(v)$ is the projection of v to U. ∎

Furthermore we have the *Gram–Schmidt* process to construct an orthonormal set,

Theorem 1.3.2: *Let us consider $\mathbb{R}^n$. Let U be generated by linearly independent set $v_1, v_2, \ldots, v_m$. Then there exist an orthogonal set $w_1, w_2, \ldots, w_m$ such that $w_i \cdot w_j = 0$ if $i \neq j$, and $w_i = v_i + \sum_{j<i} a_{ij} v_j$. Moreover let $u_i = w_i/|w_i|$. Then $\{u_1, u_2, \ldots, u_m\}$ is an orthonormal basis for U.*

---

[10]Gram, J. Danish mathematician. 1850–1916.
[11]Schmidt, E. Baltic German mathematician. 1876–1916.

Proof. We make an induction for m. For $m = 1$, let $u_1 = v_1/|v_1|$. Let us assume that inductively we have constructed $\{w_1, w_2, \ldots, w_k\}$. Then we construct $u_i = w_i/|w_i|$ for $i = 1, \ldots, k$. If there is no more v_{k+1}, then we are done. Let us assume that there is v_{k+1}. Let

$$w_{k+1} = v_{k+1} - \sum_{i=1}^{k} (v_{k+1} \cdot u_i) u_i$$

then w_{k+1} is not 0, otherwise $\{v_1, v_2, \ldots, v_{k+1}\}$ must be a linearly dependent set. It is easy to show that $\{w_1, \ldots, w_{k+1}\}$ is an orthogonal set. We take $u_{k+1} = w_{k+1}/|w_{k+1}|$. Then $|u_{k+1}| = 1$. ∎

Proposition 1.3.3: *Let us consider* $\mathbb{R}^n$. *Let* U *be generated by linearly independent* $v_1, v_2, \ldots, v_m$. *Let an orthogonal set* $w_1, w_2, \ldots, w_m$ *such that* $w_i = v_i + \sum_{j<i} a_{ij} v_j$, *and let* $w'_1, w'_2, \ldots, w'_m$ *be another orthogonal set such that* $w'_i = v_i + \sum_{j<i} a'_{ij} v_j$. *Then* $a_{ij} = a'_{ij}$ *for all* i, j, *and* $w_i = w'_i$ *for all* i.

Proof. Inductively, we have $w_1 = v_1 = w'_1$. Let us use mathematical induction, let us consider $k > 1$. Let us assume that the proposition is true for $k - 1$. It is easy to see that w_k, w'_k are both perpendicular to the subspace U_{k-1} spanned by $v_1, \ldots, v_{k-1}$. Hence the expression

$$w_k - w'_k = \sum (a_{ij} - a'_{ij}) v_k$$

which shows that $w_k - w'_k \in U_{k-1}$ and $\perp U_{k-1}$. It must be zero and must have all coefficients $a_{ij} - a'_{ij} = 0$. ∎

We have the following definition.

Definition Let us use the notations of the preceding proposition. We define the *m-dimensional volume*, $\mathbb{A}[v_1, v_2, \ldots, v_m] = \prod_j \ell$ where $\ell_j = |w_j| =$ the length of w_j. ∎

Remark: Right now, the definition of m-dimensional volume depends on the order of elements $\{v_1, v_2, \ldots, v_m\}$. Later on in Section 1.5, we will have a *formula* for the *m-dimensional volume*, $\mathbb{A}[v_1, v_2, \ldots, v_m]$, and we will give a proof to show that it is independent of the order of the vectors. ∎

Exercises

(1) Prove the n-dimensional Pythagoras theorem.

(2) In $\mathbb{R}^3$, find a vector which is perpendicular to two vectors $[1, 2, 3]^T, [1, 1, 1]^T$.

(3) Let V be a vector space with a vector x_0 and a linear fuctional y_0. Find the conditions on x_0 and y_0 such that $Ax = y_0(x)x_0$ defining a projection for all vectors $x \in V$.

(4) Find an equation satisfied by all points on the sphere of radius 1.

(5) Show that $u = [1, 1]^T$ and $v = [-1, 1]^T$ in $\mathbb{R}^2$. Find vector w such that $w \cdot v = 1, w \cdot u = 2$.

(6) In $\mathbb{R}^n$, if $v \cdot u = 0$ for all u, then $v = 0$.

1.4 The Fundamental Theorem of Matrices. Kirchhoff's Law

We will mainly discuss the *fundamental theorem* of matrices in this section. Recall that in general we have the related *first isomorphism theorem* of Noether (see the preceding Section 1.2).

We shall first prepare the ground for the fundamental theorem. Given a system of equations as follows,

$$a_{11}x_1 + a_{12}x_2 + \cdots + a_{1n}x_n = b_1$$
$$a_{21}x_1 + a_{22}x_2 + \cdots + a_{2n}x_n = b_2 \qquad (2)$$
$$\cdots\cdots$$
$$a_{m1}x_1 + a_{m2}x_2 + \cdots + a_{mn}x_n = b_m.$$

It may be rewritten as

$$Ax = b$$

where $A = (a_{ij})$ is the coefficient matrix.

For any matrix A, we have the concepts of *row space* (i.e., the subspace of K_n generated by all row vectors of A), the *column space* (i.e., the subspace of K^m generated by all column vectors of A). We have the following lemma,

Lemma 1.4.1: *An elementary row operation will not change the row space.*

Proof. Trivial. ∎

We consider the homogeneous equation $Ay = 0$. Then we have the following proposition,

Proposition 1.4.2: *Let $V = \{y : Ay = 0\} \subset K^n$. Then V is a subspace of K^n.*

Proof. Elementary linear algebra. ∎

Given an equation $Ax = b$, the equation $Ay = 0$ is called the associated *homogeneous equations*. The equation $Ax = b$ may or may not have a solution, for instance, if $A = 0$ and $b \neq 0$, then $Ax = b$ obviously has no solution. On the other hand $Ay = 0$ always has solutions, for instance, $y = 0$ is a solution. We have the following definition,

Definition Let A be an $m \times n$ matrix. Then all rows span a subspace of K_n. It is called the *row space* of A. Its dimension is called the *row-rank* of A; in symbol, *row-rank(A)*. Similarly we have the column space of A which is subspace of K^m, and the *column-rank* of A; in symbol, *column-rank(A)*. The subspace $\{y : Ay = 0\}$ is called the *nullspace* of A. Its dimension is called the *nullity* of A. Similarly, the subspace $\{y : yA = 0\}$ is called the *left-nullspace* of A. Its dimension is called the *left-nullity* of A. ∎

Proposition 1.4.3: *Let x_0 be a particular solution of $Ax = b$ and $U = $ the nullspace of A. Then x is a solution of $Ax = b$ iff $x \in (x_0 + U)$.*

Proof. If x is any solution of $Ax = b$, then $A(x - x_0) = b - b = 0$. Henceforth $x - x_0 \in U$, and $x \in x_0 + U$. On the other hand, if $x = x_0 + y$ where $y \in U$, then $Ax = Ax_0 + Ay = b$. ∎

The row operations of a matrix A naturally do not alternate the row space of A which is a subspace of K_n. We want to examine its effect on K^m. We have the following proposition,

Proposition 1.4.4: *Given a matrix $A = [c_1, c_2, \ldots, c_n]$. Let E be an elementary row operation of A with $EA = [c'_1, c'_2, \ldots, c'_n]$. Then E can be viewed as a basis change which fixes every vector in K^m.*

Proof. Let the original coordinate system be $e_1, \ldots, e_n$, where $e_i = [0, \ldots, 0, 1, 0, \ldots, 0, 0]^T$ with 1 appearing at the ith place. Then the elementary row operation of the first kind is simply interchanging the ith element in the basis with the jth element in the basis, i.e., let a column vector $c_s = [c_{1s}, \ldots, c_{ms}]^T = \sum c_{js} e_j$, and we exchange the ith coefficient and jth coefficient of c_s, we think that the coordinate system $\{e_i\}$ is changed by exchanging e_i and e_j, and any (column) vector is invariant. The elementary row operation of the second kind which multiplying the ith row by a non-zero a will change the ith e_i by multiplying it with $1/a$ and fixes all other v_j with $j \neq i$. We observe that its coordinates change accordingly. The third elementary operation is multiplying the ith row by constant b and adding the result to the jth row, we will then have

$$v = \sum b_j e_j = \sum_{s=1}^{s=i-1} b_s e_s + b_i(e_i - b e_j)$$

$$+ \sum_{s=i+1}^{s=j-1} b_s e_s + (b_j + b b_i) e_j + \sum_{s=j+1}^{s=n} b_s e_s.$$

We simply change e_i to $e'_i = e_i - b e_j$ while fixing all other $e_k = e'_k$ with $k \neq i$. It is easy to see $\{e'_j\}$ is still a coordinate system and the coordinates of the fixed v are changed to $[b_1, \ldots, b_j + b b_i, \ldots, b_m]^T$. This argument alludes to the old viewpoint of all stars moving all night and the modern viewpoint is that of the viewpoint of N. Copernicus[12] that since all stars are moving uniformly, we may think that all stars are fixed and only the coordinate system of the earth is moving. ∎

We have the following proposition,

Proposition 1.4.5: *Given a matrix $A = [r_1, r_2, \ldots, r_m]^T$. Let E be an elementary column operation of A with $AE = [r'_1, r'_2, \ldots, r'_m]$.*

---

[12]Copernicus, N. Polish mathematician and astronomer. 1473–1543.

Then $\{r_i'\}$ can be viewed as the expressions of $\{r_i\}$ with respect to a different coordinate system. In other words, an elementary column operation simply changes the coordinates of all row vectors without changing any row vector.

Now we are well-prepared to state the first part of the theorem which is named by G. Strang[13] as part of the *fundamental theorem* of linear algebra (in fact it is a misname, since it is only about matrices, in the part of finite dimensional vector spaces).

Theorem 1.4.6: *Given any matrix A. Then its row-rank = column-rank. The common number is called the rank of A.*

Proof. Let us use a sequence of elementary row operations and then use a sequence of column operations to reduce it to its normal form. Then the theorem follows from the preceding lemma and the preceding propositions. ∎

Given a matrix A in the reduced row echelon form. Those columns without a pivot are called a *free* column. We have the other part of the *fundamental theorem* of matrix as follows,

Theorem 1.4.7: *Given an $m \times n$ matrix A with nullity k and rank r. Then we have $r + k = n$.*

Proof. A row operation will not change the nullity nor the rank. It suffices to prove the case that A is in reduced row echelon form. Any column is either free or contains a pivot but not both. Hence it suffices to show the nullity of A is the number of free columns.

Since the values of the free variables can be assigned arbitrarily, we take turns to assign one of them to have the value 1 and the rest 0's. In this way we find a basis for the nullspace. Hence the number of free columns $= n - r = k$, and nullity $= k$. ∎

Note that there is no restriction on the dimension of the vector space V in the first isomorphism theorem (cf. Section 1.2). Note that

---

[13]Strang, G. American mathematician. 1934–.

in the finite dimension cases, the map ρ is given by a matrix A. Then we have $\dim(W) = $ column-rank of $A = n - $ nullity. Therefore the second part of the *fundamental theorem* follows from the *first isomorphism* and the first part of *fundamental theorem*.

The study of the equation $Ax = b$ is concentrated on the problem of finding a particular solution x_0. We have the following proposition,

Proposition 1.4.8: *The equation $Ax = b$ has a solution iff b is in the column space of A.*

Proof. Let $A = [c_1, \ldots, c_n]$ with c_i the column vectors. Let $[x_1, \ldots, x_n]^T$ be a solution of $Ax = b$. Then the equation can be written as $x_1 c_1 + \cdots + x_n c_n = b$. Therefore b is in the column space of A.

On the other hand, if b is in the column space of A, i.e., $b = x_1 c_1 + \cdots + x_n c_n$, then let $x = [x_1, \ldots, x_n]^T$, clearly, $Ax = b$, i.e., the equation has a solution. ∎

Example 1.8 (Kirchhoff's Law): In physics, we have the Ohm's law,

$$U = IR$$

where U is the difference of potential, I is the current, R is the resistance. We take the units for potential difference as volt, for current as ampere, and for resistance as ohm. Since voltage = ampere · ohm, then the units will be omitted in the formula. Furthermore, we have *Kirchhoff's laws* of networks as follows: (1) *Kirchhoff's Current Law* is that the sum of all currents flows in and flows out to a fixed point adds to zero. (2) *Kirchhoff's Voltage Law* is that the sum of all potential differences across the component involved in a loop must be zero. One of the consequences of (1) is (3) the sum of all currents flows in and flows out to a fixed part adds to zero.

In general we may show that *Kirchhoff's law*[14] follows from the fundamental theorem of matrix. For simplicity, let us only consider

---

[14]Kirchhoff, G. German physicist. 1824–1887.

an example of a network as follows,

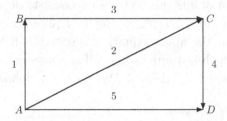

where A, B, C, D are vertices, and the directed edges are labeled 1 to 5.

Let us assume that the resistances on the ith cable is i ohm for $i = 1, \ldots, 5$. Then the following matrix represents the network

$$\mathbb{A} = \begin{bmatrix} -1/1 & 1/1 & 0 & 0 \\ -1/2 & 0 & 1/2 & 0 \\ 0 & -1/3 & 1/3 & 0 \\ 0 & 0 & -1/4 & 1/4 \\ -1/5 & 0 & 0 & 1/5 \end{bmatrix}$$

where the columns represent vertices and the rows represent the directed edges and the sign $-$ indicates the source, and the sign $+$ indicates the sink. The denominator indicates the resistance. We shall discuss the matrix $\mathbb{A}$. It is easy to see that the rank of $\mathbb{A}$ is 3, the nullity of $\mathbb{A}$ is 1 and the left-nullity is 2. The matrix $\mathbb{A}$ acts from the left on the vector space of $[v_A, v_B, v_C, v_D]^T$ where v's are the potential assigned to the vertices A, B, C, D. The matrix $\mathbb{A}$ acts from the right on the vector space of $[u_1, u_2, u_3, u_4, u_5]$ where u's are the potential differences of the directed edges 1, 2, 3, 4, 5. We have the following detailed discussions.

(1) The nullspace of $\mathbb{A}$: Clearly the 1-dimensional nullspace is,

$$N = \left\{ \begin{bmatrix} c \\ c \\ c \\ c \end{bmatrix} : c \in R \right\}.$$

The interpretation is that the number c is the potential in terms of volt. The value of potential is relative, they may be uniformly changed by a constant c.

(2) Let us consider the equation $\mathbb{A}x = b = [I_1, I_2, I_3, I_4, I_5]^T$. The interpretation of b is the vector that consists of five components with each of the current on the corresponding directed edge. We know that the above equation is solvable iff $b \in$ the column-space of $\mathbb{A}$ which is 3-dimensional. Therefore we need two general equations for five variables. We have the following system of equations,

$$I_1 - 2I_2 + 3I_3 = 0$$
$$2I_2 + 4I_4 - 5I_5 = 0.$$

Note the Ohm's law: $U = RI$, we may replace iI_i by the potential difference u_i. Then the above equation may be written as

$$u_1 - u_2 + u_3 = 0$$
$$u_2 + u_4 - u_5 = 0$$

which is precisely Kirchhoff's second law: the summation of any potential difference around a loop is zero.

(3) Let us consider the left-nullspace M of $\mathbb{A}$, $M = \{u : u\mathbb{A} = 0\}$ where u is the vector of potential difference. Then it becomes the first law of Kirchhoff: all currents flowing into a vertex should be balanced with all currents flowing out of the vertex.

(4) Let us consider a graph with all four vertices of the preceding graph connected to an outside point (say a power plant). Let us consider $y\mathbb{A} = c = [c_A, c_B, c_C, c_D]$ where c_i is the current flow into the vertex i. The vector c is in the row space of A. Therefore it is defined by

$$0 = y\mathbb{A}[1, 1, 1, 1]^T = c[1, 1, 1, 1]^T = c_A + c_B + c_C + c_D.$$

In other words, the total current flows to (or flows out from) the network must be added up to 0 which is *Kirchhoff's Law (3)*. ∎

Exercises

(1) Prove Lemma 1.4.1.

(2) Let A and B be two linear operators on an n-dimensional vector space V. If $AB = 0$; then their ranks satisfy $r(A) + r(B) \leq n$.

(3) Let V be a subspace of dimension $n - 1$ of an n-dimensional vector space U. Let ϕ be an endomorphism of U (i.e., $\phi(U) \subset U$) which fixes every element in V. Show that there is an element α such that $\phi(u) = av + t_u$, where a is common for every u and $t_u \in V$.

(4) Two matrices A, B are said to be row equivalent if there is a finite sequence of row operations which change A to B. Prove that two matrices A, B are row equivalent iff they have the same reduced row echelon form.

(5) Let A be an $m \times n$ matrix. Show that A is of rank $1 \Leftrightarrow$ there is an m-vector v (treated as $m \times 1$ matrix) and an n-vector u^T (treated as $1 \times n$ matrix) such that $A = v \cdot u^T$.

(6) Show that $b = [1, 1, 1, 1, 1]^T$ is not solvable for the matrix equation $Ax = b$, where A is the matrix in the **Example 1.8**.

(7) Let the real matrix A be given as

$$A = \begin{bmatrix} 1 & 1 & 0 & 0 \\ -2 & 0 & 2 & 0 \\ 1 & -3 & 1 & 0 \\ 1 & 0 & -1 & 4 \\ -5 & 1 & 1 & 5 \end{bmatrix}.$$

Find its rank, nullity, left-nullity, nullspace and left-nullspace.

(8) Let $P_n[x]$ be the polynomials of degree $\leq n$. Show that it is a vector space of dimension $n + 1$.

1.5 Determinant. m-Dimensional Volume in $\mathbb{R}^n$

The determinant has been discovered by a Japanese mathematician Seki Kōwa (1683). Seki Kōwa[15] studied a system of linear equations,

$$ax + by = c$$
$$dx + ey = f.$$

---

[15]Kōwa, Seki. Japanese mathematician. 1642–1708.

Certainly he wrote the above system in the traditional Chinese way, vertically, and without the modern symbols as "x", "y", "$+$", "$=$". He just wrote vertically in two columns a, b, c, d, e, f. He proved that if $ae - bd \neq 0$, then the above system always had solutions. That was the start of the determinant. Some ten years later, a great German mathematician Leibniz[16] independently discovered the theory in Europe.

Given any commutative ring R with identity $1 \neq 0$ (for definition, see Section 2.1), and an $n \times n$ matrix $A = (a_{ij})$. For simplicity, we may write $A = [c_1, c_2, \ldots, c_n]$ where $c_j = [a_{1j}, a_{2j}, \ldots, a_{nj}]^T$, and abuse the language to call c_j the jth column vector of A. In this section, we introduce the concept of determinant in the elementary way. Further discussion is postponed to Chapter 3.

Definition (Determinants by Expansions): Let $A = (a_{ij}) \in M_{n \times n}(R)$ be an $n \times n$ matrix over a ring R (in this book, a ring is always a commutative ring with identity $1 \neq 0$). Then the determinant of A, in symbol, $\det(A)$, is defined to be,

$$\det(A) = \sum \text{sign}(\sigma) a_{1i_1} \cdots a_{ni_n}$$

where the summation sign runs through all possible permutations σ of the integers $(1, 2, \ldots, n)$, and $\sigma = \left(\begin{smallmatrix} 1 & 2 & \cdot & n \\ i_1 & i_2 & \cdot & i_n \end{smallmatrix} \right)$. The function $\text{sign}(\sigma)$ is ± 1, depending on the parity of the permutation σ; if σ is *even*, that means it can be written as the product of an even number of transpositions; then it is $+1$; if it is odd, that means it can be written as the product of an odd number of transpositions, then it is -1. We may compute it by

$$\prod_{k>j} (x_k - x_j) = \text{sign}(\sigma) \prod_{k>j} (x_{\sigma(k)} - x_{\sigma(j)}). \qquad \blacksquare$$

The following proposition is trivial, while constituting one of the defining axioms (see Chapter 3),

Proposition 1.5.1: *We always have* $\det(I) = 1$.

---

[16]Leibniz, G. German polymath and philosopher. 1646–1716.

Proof. Elementary. ∎

Example 1.9: Let us consider the matrix $A = \begin{bmatrix} a_{11} & a_{12} \\ a_{21} & a_{22} \end{bmatrix} = \begin{bmatrix} 1 & 2 \\ 2 & 1 \end{bmatrix}$. Then $\det(A) = a_{11}a_{22} - a_{12}a_{21} = 1 \cdot 1 - 2 \cdot 2 = -3$. ∎

Lemma 1.5.2: *A permutation σ is even $\Leftrightarrow$ σ^{-1} is even.*

Proof. It is easy to see that $\text{sign}(\sigma\tau) = \text{sign}(\sigma) \cdot \text{sign}(\tau)$, and $\text{sign}(\sigma) \cdot \text{sign}(\sigma^{-1}) = 1$. Our lemma follows. ∎

Proposition 1.5.3: *We have $\det(A) = \det(A^T)$.*

Proof. It is clear that we have

$$\det(A) = \sum \text{sign}(\sigma)a_{1i_1} \cdots a_{ni_n}$$

and

$$\det(A^T) = \sum \text{sign}(\tau)a_{i_1 1} \cdots a_{i_n n}.$$

We have $\sigma = \begin{pmatrix} 1 & 2 & \cdot & n \\ i_1 & i_2 & \cdot & i_n \end{pmatrix}$ and $\tau = \begin{pmatrix} i_1 & i_2 & \cdot & i_n \\ 1 & 2 & \cdot & n \end{pmatrix}$. Clearly we have $\sigma\tau = 1$, therefore they have the same parity. ∎

We may denote any $n \times n$ matrix A as $[c_1, c_2, \ldots, c_n]$ where c_i is the ith column vector of the matrix A. We shall view the $\det(A)$ as $\det[c_1, c_2, \ldots, c_n]$ as a K-valued function of n vector variables. In this sense, we have the following proposition,

Proposition 1.5.4: *Given any i with $1 \leq i \leq n$, $\det[c_1, \ldots, c_i, \ldots, c_n]$ is a linear function for c_i if we fixed all other c_j, $j \neq i$.*

Proof. We have the definition of $\det A$ as

$$\det(A) = \sum \text{sign}(\sigma)a_{1i_1} \cdots a_{ni_n}.$$

We assume the summation sign runs through all possible permutations σ of the integers $(1, 2, \ldots, n)$, and $\sigma = \begin{pmatrix} 1 & 2 & \cdot & n \\ i_1 & i_2 & \cdot & i_n \end{pmatrix}$. Note that $\det$ is a linear form of any column vector $[a_{1i}, a_{2i}, \ldots, a_{ni}]^T$. Our proposition follows. ∎

Proposition 1.5.5: *Given any k, ℓ with $1 \le k < \ell \le n$, if we have $c_k = c_\ell$, then $\det[c_1, \ldots, c_i, \ldots, c_n] = 0$.*

Proof. By the previous lemma, we know that $\det(A) = \det(A^T)$. We may take c_i to be the row vectors. In the summation $\det(A) = \sum \text{sign}(\sigma) a_{1 i_1} \cdots a_{n i_n}$, for any pair $r < s$, we pair the permutations σ, γ with $\sigma(i) = \sigma(i)$ for all $i \ne k, \ell$, and $\sigma(k) = \gamma(\ell) = r$, $\sigma(\ell) = s = \gamma(k)$. Since $c_k = c_\ell$, it is easy to see that the two terms corresponding to σ, γ have the same absolute value. While letting $\tau = \sigma \gamma^{-1}$, then we have

$$\tau = \begin{pmatrix} 1 & 2 & k & \cdot & \ell & \cdot & n \\ 1 & 2 & \ell & \cdot & k & \cdot & n \end{pmatrix}.$$

It is easy to see that $\text{sign}(\tau) = -1$ and the sum of the pair is zero. Our proposition is proved. ∎

Given a matrix A, the computation of $A^T A$ is useful and appears many times in the *least square approximation, singular value decomposition* and others. In the following example, we will indicate another usage of $A^T A$ for the computations of m-dimensional volume in $\mathbb{R}^n$ with the usual dot product to define the length.

Example 1.10 (Length): Let us consider the usual length $|v|$ of a vector $v = [a_1, a_2, \ldots, a_n]^T = A$. Then we have the *generalized Pythagoras theorem,*

$$|v|^2 = \det(A^T A) = \left(\sum_i a_i^2 \right),$$

where the last sum means that the summation is over all subsets s of $\{1, \ldots, n\}$ with s singleton and $A_s = $ the sth component of A.

Area: Let us consider the real space $\mathbb{R}^n$ with the usual definition of length and the parallelogram spanned by two linearly vectors $v = [a_1, \ldots, a_n]^T, u = [b_1, \ldots, b_n]^T$. Let $A = [v, u]$ be $n \times 2$ matrix and the area of it be $\mathbb{A}[v, u]$. Then we have

$$\mathbb{A}^2 = |v|^2 |u|^2 (\sin \theta)^2 = |v|^2 |u|^2 - (v \cdot u)^2 = (\det A^T A).$$

The n-dimensional volume: Let us consider the real space $\mathbb{R}^n$ with the usual definition of length and the parallelopiped spanned by n linearly vectors $v_1 = [a_{11}, \ldots, a_{n1}]^T, \ldots, v_n = [a_{1n}, \ldots, a_{nn}]^T$. Let $A = [v_1, \ldots, v_n]$ be $n \times n$ matrix and the volume of it be $\mathbb{A}[v_1, \ldots, v_n]$. Then we have

$$\mathbb{A}^2 = (\det A[v_1, \ldots, v_n])^2 = (\det A^T A). \qquad \blacksquare$$

Let us consider the real space $\mathbb{R}^n$ with the usual definition of length and the *parallelopiped* spanned by m linearly vectors $v_1 = [a_{11}, \ldots, a_{n1}]^T, \ldots, v_m = [a_{1m}, \ldots, a_{nm}]^T$. Let $A = [v_1, \ldots, v_m]$ be an $n \times m$ matrix and the volume of the *parallelopiped* be $\mathbb{A} = \mathbb{A}[v_1, \ldots, v_m]$. We shall only consider a **special case** to make the reasoning simple.

Proposition 1.5.6: *Let us assume that the vectors $\{v_1, \ldots, v_m\}$ are perpendicular, i.e., $v_i \cdot v_j = 0$ if $i \neq j$. Let us assume that $\ell_i = |v_i|$. Then we have $(\mathbb{A}[v_1, \ldots, v_m])^2 = \det(A^T A)$.*

Proof. We have

$$\det(A^T A) = \det \begin{bmatrix} \ell_1^2 & 0 & 0 & 0 \\ 0 & \ell_2^2 & 0 & 0 \\ & . & . & . \\ 0 & 0 & 0 & \ell_m^2 \end{bmatrix} = \prod_i \ell_i^2 = \mathbb{A}^2. \qquad \blacksquare$$

We push the computation one step further. Consider linearly independent vectors $v_1, \ldots, v_m$, it follows from the Gram–Schmidt theorem of Section 1.2 that we can find vectors $w_1, \ldots, w_m$ such that (1) $w_i \cdot w_j = 0$ for all $i \neq j$, (2) $w_1 = v_1, \ldots, w_i = v_i + \sum_{j<i} a_{ij} v_j, \ldots$. Therefore it is easy to see that we may reach $w_1, \ldots, w_m$ from $v_1, \ldots, v_m$ by finitely many transformations of the form $u_k = v_k$ for $k \neq i$ and $u_i = v_i + a v_j$ where $j < i$.

Proposition 1.5.7: *Let $v_1, \ldots, v_m$ be linearly independent vectors in $\mathbb{R}^n$ and for some fixed i, $u_k = v_k$ for $k \neq i$ and $u_i = v_i + a v_j$ where $j < i$. Let $A = [v_1, \ldots, v_m]$ and $B = [u_1, \ldots, u_m]$. Then we have $\det(A^T A) = \det(B^T B)$.*

Proof. We have

$$\det(A^T A) = \det \begin{bmatrix} v_1 \cdot v_1 & \cdots & v_1 \cdot v_i & \cdots & v_1 \cdot v_m \\ \vdots & \cdots & \vdots & \cdots & \vdots \\ v_i \cdot v_1 & \cdots & v_i \cdot v_i & \cdots & v_i \cdot v_m \\ \vdots & \cdots & \vdots & \cdots & \vdots \\ v_m \cdot v_1 & \cdots & v_m \cdot v_i & \cdots & v_m \cdot v_m \end{bmatrix}$$

and

$$\det(B^T B) = \det \begin{bmatrix} u_1 \cdot u_1 & \cdots & u_1 \cdot u_i & \cdots & u_1 \cdot u_m \\ \vdots & \cdots & \vdots & \cdots & \vdots \\ u_i \cdot u_1 & \cdots & u_i \cdot u_i & \cdots & u_i \cdot u_m \\ \vdots & \cdots & \vdots & \cdots & \vdots \\ u_m \cdot u_1 & \cdots & u_m \cdot u_i & \cdots & u_m \cdot u_m \end{bmatrix}.$$

Let us make the substitution $u_i = v_i + av_j, u_k = v_k \,\forall k \neq i$. In the resulting matrix, we multiply the jth row with $-a$, add the result to the ith row, and multiply the jth column by $-a$, and add the result to the ith column. We get the expansion of $(A^T A)$, and we conclude $\det(A^T A) = \det(B^T B)$. ∎

According to the Gram–Schmidt theorem, a sequence of finitely many transformations of the type in the previous proposition will transform any linearly independent set $[v_1, \ldots, v_m]$ to a mutually perpendicular set $[w_1, \ldots, w_m]$, hence the previous proposition shows that $\det(A^T A)$ will be the square of the m-dimensional volume of the fixed ordered vectors $\{v_1, \ldots, v_m\}$. So far we have fixed the order of the vectors $v_1, \ldots, v_m$, now we want to change the order to any one, and **claim** that $\mathbb{A}([v_1, \cdot, v_m])$ is the same. To prove that, we have to show $\mathbb{A}([v_1, \cdot, v_m])$ is invariant with respect to any interchange of i, j.

Proposition 1.5.8: *Let $u_k = v_k$ for $k \neq i, j$ and $u_i = v_j, u_j = v_i$ where $j \neq i$. Let $A = [v_1, \ldots, v_m]$ and $B = [u_1, \ldots, u_m]$. Then we have $\det(A^T A) = \det(B^T B)$.*

Proof. We have

$$\det(A^T A) = \det \begin{bmatrix} v_1 \cdot v_1 & \cdots & v_1 \cdot v_j & \cdots & v_1 \cdot v_i & \cdots & v_1 \cdot v_m \\ \vdots & \cdots & \vdots & \cdots & \vdots & \cdots & \vdots \\ v_j \cdot v_1 & \cdots & v_j \cdot v_j & \cdots & v_j \cdot v_i & \cdots & v_j \cdot v_m \\ \vdots & \cdots & \vdots & \cdots & \vdots & \cdots & \vdots \\ v_i \cdot v_1 & \cdots & v_i \cdot v_j & \cdots & v_i \cdot v_i & \cdots & v_i \cdot v_m \\ \vdots & \cdots & \vdots & \cdots & \vdots & \cdots & \vdots \\ v_m \cdot v_1 & \cdots & v_m \cdot v_j & \cdots & v_m \cdot v_i & \cdots & v_m \cdot v_m \end{bmatrix}$$

and

$$\det(B^T B) = \det \begin{bmatrix} v_1 \cdot v_1 & \cdots & v_1 \cdot v_i & \cdots & v_1 \cdot v_j & \cdots & v_1 \cdot v_m \\ \vdots & \cdots & \vdots & \cdots & \vdots & \cdots & \vdots \\ v_i \cdot v_1 & \cdots & v_i \cdot v_i & \cdots & v_i \cdot v_j & \cdots & v_i \cdot v_m \\ \vdots & \cdots & \vdots & \cdots & \vdots & \cdots & \vdots \\ v_j \cdot v_1 & \cdots & v_j \cdot v_i & \cdots & v_j \cdot v_j & \cdots & v_j \cdot v_m \\ \vdots & \cdots & \vdots & \cdots & \vdots & \cdots & \vdots \\ v_m \cdot v_1 & \cdots & v_m \cdot v_i & \cdots & v_m \cdot v_j & \cdots & v_m \cdot v_m \end{bmatrix}.$$

We interchange ith, jth rows and ith, jth columns, then we have $\det(A^T A) = \det(B^T B)$. ∎

We have the following theorem,

Theorem 1.5.9: *We always have* $\det(A^T A) = \mathbb{A}^2$. ∎

Exercises

(1) Find all operators ϕ on $\mathbb{R}^2$ which only fix the origin 0.
(2) Find the 2-dimensional volume of the parallelogram spanned by vectors $[1, 1, 2]^T$, $[2, 3, 4]^T$.
(3) Find the 3-dimensional volume of the following parallelopiped in the real 4-dimensional space $\mathbb{R}^4$ spanned by

$$\begin{bmatrix} 0 \\ 1 \\ 2 \\ 3 \end{bmatrix}, \begin{bmatrix} 3 \\ 2 \\ 1 \\ 0 \end{bmatrix}, \begin{bmatrix} 2 \\ 3 \\ 0 \\ 1 \end{bmatrix}.$$

(4) Find the 3-dimensional volume of the following parallelopiped in the real 4-dimensional space $\mathbb{R}^4$ spanned by

$$\begin{bmatrix} 0 \\ 1 \\ 2 \\ 3 \end{bmatrix}, \quad \begin{bmatrix} 1 \\ 1 \\ 1 \\ 0 \end{bmatrix}, \quad \begin{bmatrix} 5 \\ 3 \\ 0 \\ 1 \end{bmatrix}.$$

(5) Show that with $d(n)$ defined as the determinant of the following $n \times n$ matrix over $\mathbb{R}$,

$$\det \begin{bmatrix} 1 & 2 & 0 & \cdot & \cdot & 0 \\ 2 & 1 & 2 & 0 & \cdot & 0 \\ 0 & 2 & 1 & 2 & \cdot & 0 \\ \cdot & \cdot & \cdot & \cdot & \cdot & \cdot \\ 0 & \cdot & 0 & 2 & 1 & 2 \\ 0 & \cdot & \cdot & 0 & 2 & 1 \end{bmatrix}$$

where $n \geq 1$. We define $d(0) = 1$, $d(-1) = 0$. Show that $d(n) = d(n-1) - 4d(n-2)$ for all $n \geq 2$.

1.6 Dimension Theory of General Vector Spaces

We want to prove the following propositions in this section,

Proposition 1.6.1: *Let V be a vector space. Then the cardinality of a generating set S is greater than or equal to the cardinality of a linearly independent set S'.* ∎

Proposition 1.6.2: *Any two bases A, B of V must have the same cardinality.* ∎

Preliminary Materials

Definition Let V be a vector space, and A, B two subsets of V. We say that A, B are linearly independent iff we have

$$\sum_{finite} \alpha_i a_i + \sum_{finite} \beta_j b_j = 0$$

where $a_i \in A, b_j \in B$, then $\alpha_i = \beta_j = 0$ for all i, j. Especially, if A, B are linearly independent, then $A \cap B = \emptyset$. ∎

We have the following lemma,

Lemma 1.6.3: *Let V be a vector space, and A, B two subsets of V. Then A, B are linearly independent iff* (1) $A \cap B = \emptyset$. (2) $A \cup B$ *is a linearly independent set.*

Proof. Exercises. ∎

The Proof of the First Proposition

We use Zorn's lemma (cf. Section 1.1). For these purposes, we (1) first define a family F and show that it is not empty, (2) then define a partial ordering $\leq$ on F, (3) further establish that the hypothesis of Zorn's lemma is satisfied, (4) therefore we use the conclusion of Zorn's lemma, i.e., the existence of a maximal element in F, (5) use the existence of a maximal element to prove the proposition. We carry out the above five steps in the following,

(1) Let S be the generating set, S' be the linearly independent set to be compared in Proposition 1.6.1. Let $W = S \cap S'$, and we define the family F as $F = \{(T, \rho, T')\}$ where

- $W \subset T \subset S, W \subset T' \subset S'$, and ρ is a bijective map : $T \mapsto T'$ and the restrictive of ρ to W is the identity map.
- The two sets $T, S' \backslash T'$ are linearly independent.

Notice that $(W, 1_w, W) \in F$. Even if $W = \emptyset$, the element $(W, 1_w, W)$ still exists. Therefore $F \neq \emptyset$.

(2) We define a partial ordering $\leq$ as follows. We define $(T_i, \rho_i, T_i') \leq (T_j, \rho_j, T_j')$ iff

- $T_i \subset T_j, T_i' \subset T_j'$.
- The restriction of ρ_j to T_i is ρ_i.

It is easy to see that $\leq$ is a partial ordering.

(3) We want to show that the hypothesis of Zorn's lemma is satisfied. Let $\mathbb{G} = \{(T_i, \rho_i, T_i')\}$ be a linear subfamily of F. We want to show that it has an upper bound.

Let $T = \cup T_i$, $T' = \cup T_i'$, and ρ is defined by $\rho(t) = \rho_i(t)$ if $t \in T_i$. It is easy to see that ρ is well-defined and bijective. We want to show that $T, S'\backslash T'$ are linearly independent. According to the lemma, we have to prove two things: (a) $T \cap (S'\backslash T') = \emptyset$ and (b) $T \cup (S'\backslash T')$ is a linearly independent set.

Let us prove the statement: $T \cap (S'\backslash T') = \emptyset$. If not, let $t \in T \cap (S'\backslash T')$, then we have $t \in T = \cup T_i$. Therefore we have $t \in T_j$ for some j. We have

$$t \in T_j \cap (S'\backslash T') \subset T_j \cap (S'\backslash T_j) = \emptyset.$$

A contradiction!

Let us prove the statement: $T \cup (S'\backslash T')$ is a linearly independent set. Suppose not. Then there is a non-trivial relation among its elements t_i,

$$\sum a_i t_i = 0.$$

To save our notations, we may rearrange the indices i such that $t_1, \ldots, t_s \in T$ and $t_{s+1}, \ldots, t_m \in S'\backslash T'$. It is not hard to see that there is an r with $t_1, \ldots, t_s \in T_r$ and $t_{s+1}, \ldots, t_m \in S'\backslash T' \subset S'\backslash T_r'$. However $T_r \cup (S'\backslash T_r')$ is a linearly independent set. Contradiction!

We conclude that $(T, \rho, T') \in F$ and (T, ρ, T') is an upper bound of the linear subfamily.

(4) Therefore we shall use the conclusion of Zorn's lemma, i.e., the existence of a maximal element, $(\bar{T}, \bar{\rho}, \bar{T}')$, in F.

(5) We **claim** that $\bar{T}' = S'$. If the claim is true, then a subset $\bar{T} \subset S$ will have the same cardinality with S'. Therefore we have $\text{card}(S) \geq \text{card}(S')$. Our proposition is thus proved. The proof of the **claim** is as follows.

We assume that $S'\backslash\bar{T}' \neq \emptyset$, then we want to deduce that $(\bar{T}, \bar{\rho}, \bar{T}')$ is not a maximal element in F. This will be the contradiction we are looking for.

If $S'\backslash\bar{T}' \neq \emptyset$, let $s' \in S'\backslash\bar{T}'$. Since $\bar{T} \cup (S'\backslash\bar{T}')$ is a linearly independent set, therefore the subspace $\langle \bar{T} \rangle$ spanned by $\bar{T}$ is not the whole space V. It means that $\langle \bar{T} \rangle$ does not contain S which is a generating set of V. There is an element $s \in S$ such that

$$s \notin \langle \bar{T} \rangle, \ s \notin \bar{T}, \ \{s\} \cup \bar{T} \text{ is a linearly independent set.}$$

We have two possibilities. Case (A). The two sets $\{s\} \cup \bar{T}$ and $S' \backslash \bar{T}'$ are linearly independent. Case (B). The two sets $\{s\} \cup \bar{T}$ and $S' \backslash \bar{T}'$ are linearly dependent.

Case (A). This is an easy case. We may just use s' without bothering to modify it, and define $T^* = \{s\} \cup \bar{T}$, $(T^*)' = \{s'\} \cup \bar{T}'$, and define the map ρ^* as

$$\rho^*(t) = \bar{\rho}(t), \text{ if } t \neq s, \ \rho^*(t) = s', \text{ if } t = s.$$

It is easy to see that $(T^*, \rho^*, (T^*)') \in F$, and

$$(\bar{T}, \bar{\rho}, \bar{T}') < (T^*, \rho^*, (T^*)')$$

hence $(\bar{T}, \bar{\rho}, \bar{T}')$ is not a maximal element in F. A contradiction.

Case (B). There are two further cases. Subcase (B1): the two sets $\{s\} \cup \bar{T}$ and $S' \backslash \bar{T}'$ have a non-empty intersection. Subcase (B2): the two sets $\{s\} \cup \bar{T}$ and $S' \backslash \bar{T}'$ have an empty intersection, while the set $\{s\} \cup \bar{T} \cup S' \backslash \bar{T}'$ is a linearly dependent set (otherwise, it belongs to Case (A)).

Subcase (B1). Since the set $\{s\} \cup \bar{T}$ and $S' \backslash \bar{T}'$ have a non-empty intersection, the intersection must be $\{s\}$. Therefore we have $s \in S' \backslash \bar{T}'$. We shall define $T^* = \{s\} \cup \bar{T}$, $(T^*)' = \{s\} \cup \bar{T}'$, and define the map ρ^* as

$$\rho^*(t) = \bar{\rho}(t), \text{ if } t \neq s, \ \rho^*(t) = s, \text{ if } t = s.$$

It is easy to see that $(T^*, \rho^*, (T^*)') \in F$, and

$$(\bar{T}, \bar{\rho}, \bar{T}') < (T^*, \rho^*, (T^*)')$$

hence $(\bar{T}, \bar{\rho}, \bar{T}')$ is not a maximal element in F. A contradiction.

Subcase (B2). The two sets $\{s\} \cup \bar{T}$ and $S' \backslash \bar{T}'$ have an empty intersection, while the set $\{s\} \cup \bar{T} \cup S' \backslash \bar{T}'$ is a linearly dependent set. Therefore there is a relation

$$as + \underset{finite}{\sum} a_i t_i + \underset{finite}{\sum} b_j t'_j = 0, \quad t_i \in \bar{T}, \ t'_j \in S' \backslash \bar{T}'.$$

Note that $a \neq 0$. Otherwise it will be a relation among elements in $\bar{T} \cup (S' \backslash \bar{T}')$ which is known to be a set of linearly independent elements. Furthermore, we may change a to -1. Then the equation is unique. Otherwise, we may subtract one from the other, and get a non-trivial equation without the term s. We notice that not all coefficients $b_j = 0$. Otherwise it contradicts the property that the set $\{s\} \cup \bar{T}$ is a linearly independent set. We may assume that $b_1 \neq 0$. We shall define $T^* = \{s\} \cup \bar{T}$, $(T^*)' = \{t_1'\} \cup \bar{T}'$, and define the map ρ^* as

$$\rho^*(t) = \bar{\rho}(t), \text{ if } t \neq s, \ \rho^*(t) = t_1', \text{ if } t = s.$$

It is easy to see that $(T^*, \rho^*, (T^*)') \in F$, and

$$(\bar{T}, \bar{\rho}, \bar{T}') < (T^*, \rho^*, (T^*)')$$

hence $(\bar{T}, \bar{\rho}, \bar{T}')$ is not a maximal element in F. A contradiction.

If we assume $\bar{T}' \neq S'$, we always get a contradiction. Henceforth, we conclude that $\bar{T}' = S'$. Thus we have

$$Card(S) \geq Card(\bar{T}) = Card(\bar{T}') = Card(S'). \qquad \blacksquare$$

We shall use the preceding proposition to prove the following proposition,

Proposition: *Any two bases of V will have the same cardinality.*

Proof. Let A, B be two bases of V. We view A as a generating set and B as a linearly independent set. It follows from the preceding proposition that

$$Card(A) \geq Card(B).$$

We may view A as a linearly independent set and B as a generating set, then we have

$$Card(A) \leq Card(B).$$

It follows from the set theory that

$$Card(A) = Card(B). \qquad \blacksquare$$

Exercises

(1) Let V be any non-zero vector space. Show that $\oplus_1^\infty V$ and $\prod_1^\infty V$ are of different dimensions.

(2) Show that $\oplus_1^n \mathbb{R}$ and $\mathbb{R}$ are of different dimensions and of the same cardinality.

(3) Let V be any vector space, then there is a vector space U over the same field K with bigger dimension.

(4) Find the dimension of $M_{n \times m}(K)$ as an additive vector space over K.

(5) Finish the proof of the Lemma 1.6.3 of this section.

Chapter 2

Module

The main theorem in this chapter is the theorem on the Smith[1] Normal Form for a matrix over a P.I.D.

2.1 Rings

We wish to generalize the *normal form* of a matrix over a field. Soon we discover that only after suitable generalization we may understand the subject better. We have the following generalization of the concept of *field* to a *ring*.[2] It is defined as follows,

Definition (Ring): A non-empty set K together with two operations $+, \times$ (we may say $(K, +, \times)$) form a *ring* if

(1) The set K is a commutative group with respect to $+$. We shall name the unit element of addition as 0.

(2) The set $K \backslash \{0\}$ is a semi-group with respect to $\times$. If there is a unit, we denote the unit as 1. We further define $k \times 0 = 0 \times k = 0$, for all $k \in K$.

(3) The relations between the two operations $+$ and $\times$ are distributive i.e., $k_1 \times (k_2 + k_3) = k_1 \times k_2 + k_1 \times k_3$ and $(k_2 + k_3) \times k_1 = k_2 \times k_1 + k_3 \times k_1$ for all $k_i \in K$. ∎

---

[1]Smith, H. J. Irish mathematician. 1826–1883.
[2]Both field and ring are defined by Dedekind.

Definition A ring is called a *domain* if there is no non-zero zero-divisors, i.e., if $ab = 0$ and $a \neq 0$, then $b = 0$, $b \neq 0$, then $a = 0$. A subset I of R is said to be an ideal iff (1) $i, j \in I$ implies $i + j \in I$. (2) For all elements $i \in I$ and $r \in R$, we always have $ri \in I$ and $ir \in I$. A ring is a principal ideal ring iff every ideal I is principal i.e., $I = (a)$ for some $a \in R$. A ring is said to be a P.I.D. if it is a principal ideal ring and a domain. ∎

Remark: Instead of writing the ring multiplication as $\times$, we shall denote it by "$\cdot$". The simplest ring is the zero ring that consists of the element 0 only. If we assume that the multiplication is *commutative*, we shall call it a *commutative ring*. If further there is a multiplicative unity $1 \neq 0$, then we shall call it a *commutative ring with identity*. In this book, we shall assume every ring used is a commutative ring with identity if it is not stated otherwise.

If our ring is a ring without identity (for instance, consider the ring of all even numbers), we may follow Jacobson and call the objects thus defined *rng* (a rng is a ring without i). ∎

Example 2.1: Every field is a ring, while not every ring is a field. The most important rings which are not fields are $\mathbb{Z}$, the ring of integers, and $K[x_1, \ldots, x_n] = \{f(x_1, \ldots, x_n) : f(x_1, \ldots, x_n)$ is a polynomial in n-variables $x_1, \ldots, x_n\}$. We may identify the variables $x_1, \ldots, x_n$ with coordinates in A_K^n and call $K[x_1, \ldots, x_n]$ the polynomial ring of n variables over A_K^n. Note that polynomials are everywhere defined and we may think of $K[x_1, \ldots, x_n]$ as the set of polynomials on A_K^n. Similarly, we may think that $\mathbb{Z}$ is the set of globally-defined functions on the set of all prime ideals $\{p\}$ of $\mathbb{Z}$ with the *value* $f(p) = f \bmod p \in Z/pZ$ where $f \in Z$. ∎

A ring with identity is very different from a ring without identity. To understand this point, we have to have some knowledge of *set theory*. We assume that we have the knowledge of Section 1.1 about the axiom system of set theory, and that the axiom of choice may be replaced by *Zorn's lemma* in the usual system of axioms of set.

One of the most important concept in the theory of ring is the ideal which was invented by Kummer[3] for the study of *Fermat's last problem*. It turns out to be an important concept in algebra while insignificant in solving Fermat's last problem. We have,

Definition Let R be a ring (commutative ring with identity). An ideal I is said to be a maximal ideal if it is maximal among all proper ideals with respect to inclusion. ∎

Remark: If we do not require the multiplication to be commutative as in this book, then we have the concepts of left-ideal, right-ideal, two-sided ideal. ∎

We have the following proposition,

Proposition 2.1.1: *Let R be a ring (in this book, ring means a commutative ring with identity $1 \neq 0$, if we do not state otherwise). Then there is a maximal ideal I in R.*

Proof. We want to use *Zorn's lemma*. We have to verify the hypothesis in Zorn's lemma. Let $F = \{I : I \text{ ideal of } R, \text{ and } 1 \notin I\}$. We use the inclusion as the inequality. Let $\{I_i; i \in S\}$ be a linear ordered subfamily of F. Let $I = \cup_i I_i$. We claim that I is an ideal.

(1) Let $a, b \in I$. By the definition of I, we have $a \in I_i$, $b \in I_j$ for some indices i, j. Since $\{I_i\}$ is a linear ordered subfamily. We must have either $I_i \subset I_j$ or $I_j \subset I_i$. Without losing generality, we may assume $I_i \subset I_j$. Therefore $a, b \in I_j$ which is an ideal, and $a + b \in I_j \subset I$. It is easy to see that I is an additive group, and $aI \subset I$. Henceforth I is an ideal.

(2) We claim $1 \notin I$. Since $1 \in I = \cup_i I_i$, which implies that $1 \in I_j$ for some j. This is impossible.

By (1) and (2), we conclude that $I \in F$. Therefore I is an upper bound for the linear ordered subfamily. Using *Zorn's lemma*, there is a maximal ideal for R. ∎

---

[3]Kummer, E. German applied mathematician. 1810–1893.

Example 2.2: We will give an example of a rng (ring without i) which does not have a maximal ideal. Let the commutative additive group G be defined as $G = \{a/p^n : a, n, p \text{ integers}, p \text{ fixed prime}\}$. We further define the product $a \times b = 0$ always. Then it is easy to see that it forms a rng, and we use the same name G for the rng. Let $G_n = \{a/p^n : a \in \mathbb{Z}\}$. Then it is easy to see that G_n is an ideal of G. Moreover $G = \cup_n G_n$. It is easy to see that there is no maximal subgroup, hence as a rng, there is no maximal ideal. ∎

Example 2.3: Let us consider the ring of all m-differentiable functions $D^m(\aleph) = \{$all functions with all mth partial derivatives over the domain $\aleph\}$. Then clearly $D^m(\aleph)$ is a ring. ∎

Example 2.4: Let us consider the ring $\mathbb{Z}/4\mathbb{Z} = \{0, 1, 2, 3\}$. Then it has only three ideals $I_0 = (0), I_1 = (0, 1, 3) = (1) = (3), I_2 = (0, 2) = (2)$. So while it is a principal ideal ring, we see that $2 \cdot 2 = 0, 2 \neq 0$ and so it is not a domain and not a P.I.D. ∎

The most important P.I.D.'s in this book are integers $\mathbb{Z}$ and $K[x]$ where K is a field. We have the following discussions.

Example 2.5: It is clear that $\mathbb{Z}$ is a domain. We assume the following *Euclidean[4] Algorithm*: given any $a, b \in \mathbb{Z}$, with $b \neq 0$, then there exist unique q, r with $0 \leq r < |b|$ such that $a = q \cdot b + r$. Let I be any ideal of $\mathbb{Z}$. If $I = (0)$ is the zero ideal of $\mathbb{Z}$, then I is certainly principal. Otherwise, there is a non-zero element $a \in I$. We know $-a \in I$, therefore there must be a positive element in I. Let b be the smallest positive element in I. Then for any $a \in I$, by the Euclidean Algorithm, we must have q, r such that

$$a = q \cdot b + r, \quad 0 \leq r < |b| = b.$$

Clearly $r \in I$, hence $r = 0$ and $b|a$, i.e., $I = (b)$. This shows that $\mathbb{Z}$ is a P.I.D. ∎

---

[4]Euclid. Greek mathematician from Alexandria. Mid-4th century BC to mid-3rd century BC.

Example 2.6: Let p be a prime number and $\mathbb{Z}_p$ be the *localization* of $\mathbb{Z}$ at the prime ideal (p), i.e., $\mathbb{Z}_p = \{\frac{a}{b} : a, b \in \mathbb{Z}, b \notin (p)\}$. Then it is easy to see that $\mathbb{Z}_p$ is a P.I.D. (See the Exercises.)

We may consider the *p-adic topology* in $\mathbb{Q}$ as follows: let $r, s \in \mathbb{Z}_p$ with $r - s = \frac{a}{b}$ and $b \notin (p)$ with $p^n | a, p^{n+1} \nmid a$ where $n = \infty$ if $a = 0$. Then we define $d_p(r, s) = p^{-n}$. We know the function d_p thus defined is a distance. (See the Exercises.) Using the concept of *Cauchy*[5] sequences, we will find the completion of $\mathbb{Q}$, $\hat{\mathbb{Q}}_p$, and name it the *p-adic numbers* $\hat{\mathbb{Q}}_p$. It is a field.

In $\hat{\mathbb{Q}}_p$, let α, β, γ be three numbers, say, $d_p(\alpha, \beta) = r$ and $d_p(\alpha, \gamma) < r$, geometrically, we use α as the center to draw a circle with radius r, then γ is an interior point. We conclude that $d_p(\beta, \gamma) = r$ (see the Exercises). In other words, every interior point of a circle is a center. By the same argument, we can show that every triangle is isosceles (see the Exercises). ∎

Example 2.7: Let K be a field. It is clear that $K[x]$ is a domain. We assume the following *Euclidean Algorithm*: given any $f(x), g(x) \in K[x]$, with $g(x) \neq 0$, let $\deg(f(x))$ be the degree of $f(x)$ in x (recall $\deg(0) = -\infty$). Then there exist unique $q(x), r(x)$ with $\deg(r(x)) < \deg(g(x))$ such that

$$f(x) = q(x) \cdot g(x) + r(x).$$

Let I be any ideal of $K[x]$. If $I = (0)$, the zero ideal, then I is certainly principal. Otherwise, there is a non-zero element $f(x) \in I$. Let $g(x)$ be an element in I with $\deg(g(x))$ minimal. Then for any $f(x) \in I$, by the Euclidean Algorithm, we must have $q(x), r(x)$ such that

$$f(x) = q(x) \cdot g(x) + r(x), \quad \deg(r(x)) < \deg(g(x)).$$

Clearly $r(x) \in I$, hence $r(x) = 0$ and $g(x) | f(x)$, i.e., $I = (g(x))$. This shows that $K[x]$ is a P.I.D. ∎

---

[5]Cauchy, A. French mathematician, engineer and physicist. 1789–1857.

Example 2.8 (Finite Fields): Let F_p be the prime field of p elements where p is a prime number, i.e., $F_p = \mathbb{Z}/p\mathbb{Z}$. Let $f(x)$ be an *irreducible polynomial* $\in F_p[x]$ with degree n. Then $F_p[x]/(f(x))$ is a finite field. Moreover all finite fields are constructed in this way.

It is known that all finite fields of p^n elements are isomorphic, so we may use $GF(p^n)$ (i.e., Galois field of p^n elements) to denote any one of them. All non-zero elements of $GF(p^n)$ form a cyclic multiplicative group of order $N = p^n - 1$.

The interesting cases are $p = 2$ since the finite field of 2^n elements can be easily simulated in a computer. It is easy to see that the elements in this field are of the forms $a_0 + a_1\bar{x} + \cdots + a_{n-1}\bar{x}^{n-1}$ where $a_i \in \mathbb{Z}_2$ and

$$(a_0 + a_1\bar{x} + \cdots + a_{n-1}\bar{x}^{n-1}) + (b_0 + b_1\bar{x} + \cdots + b_{n-1}\bar{x}^{n-1})$$

$$= (a_0 + b_0) + (a_1 + b_1)\bar{x} + \cdots + (a_{n-1} + b_{n-1})\bar{x}^{n-1}$$

$$(a_0 + a_1\bar{x} + \cdots + a_{n-1}\bar{x}^{n-1}) \cdot (b_0 + b_1\bar{x} + \cdots + b_{n-1}\bar{x}^{n-1})$$

$$= \left(\sum_k \sum_{i+j=k} a_i b_j \bar{x}^k \bmod (f(x)) \right).$$

We shall take n bits to record $a_1 a_2 \cdots a_{n-1}$ for the field elements. The addition of the field elements is routine, while the multiplication is produced by a module operation. For n reasonably small, we may form a *multiplication table* in the computer, and do multiplication via a look-up in the table. ∎

Exercises

(1) Show that $\mathbb{Q}[i]$ is a field, where $\mathbb{Q}$ is the field of rational numbers, i is a complex number such that $i^2 - 1$ and $\mathbb{Q}[i]$ is the set of all complex numbers which can be written as $a + bi$ for a, b rational numbers.

(2) Show that $K[[x]]$ is a ring where K is a field, x is a symbol (i.e., a variable) and $K[[x]]$ is the power series ring $= \{\sum_0^\infty a_n x^n : a_n \in K\}$ under the usual sum and multiplication.

(3) Show that $K((x))$ is a field where K is a field, x is a symbol and $K((x))$ is the formal meromorphic function field $\{\sum_m^\infty a_n x^n : m \in \mathbb{Z}, a_n \in K\}$ under the usual sum and multiplication.

(4) Let p be a prime number and $\mathbb{Z}_p$ be the *localization* of $\mathbb{Z}$ at the prime ideal (p). Show that $\mathbb{Z}_p$ is a P.I.D.

(5) Write down a multiplication table for a field of eight elements.

(6) Show that the *p-adic numbers* $\hat{\mathbb{Q}}_p$ is a field.

(7) Show that the non-negative function d_p in Example 2.8 defines a distance, i.e., (1) $d_p(r,s) = 0 \Longleftrightarrow r = s$, (2) $d_p(r,s) = d_p(s,r)$, (3) $d_p(r,s) + d_p(e,t) \le d_p(r,t)$.

(8) In $\hat{\mathbb{Q}}_p$, let α, β, γ be three numbers, say, $d_p(\alpha, \beta) = r$ and $d_p(\alpha, \gamma) < r$. Show that $d_p(\beta, \gamma) = r$. In other words, every interior point of a circle is a center. By the same argument, show that every triangle is isosceles.

(9) Show that every finite field F is with cardinality p^n, where p is a prime number.

(10) Let F be a finite field. Show that $F \backslash \{0\}$ is a cyclic multiplicative group.

2.2 Modules

One of the most useful generalizations of vector spaces is the concept of *modules* which is defined by Dedekind as

Definition (Module): Let R be a commutative ring with identity which will be called *scalar ring*. A non-empty set M together with two operations $+, \cdot$ (we may say $(M, +, \cdot)$) forms a *module* over R, or called R-module, if

(1) The set M is a commutative group with respect to $+$. We shall name the unit element as 0.

(2) The *scalar multiplication* is defined, i.e., for any $m \in M$ and any $r \in R$ the multiplication $r \cdot m$ is defined and is in M.

(3) The relations between the two operations $+$ and $\cdot$ are associative and distributive, i.e., $(k_1 \cdot k_2) \cdot v = k_1 \cdot (k_2 \cdot v)$,

$k \cdot (v_1 + v_2) = k \cdot v_1 + k \cdot v_2$ for all $k \in K$ and $v_1, v_2 \in V$ and
$(k_1 + k_2) \cdot v = k_1 \cdot v + k_2 \cdot v$ for all $k_1, k_2 \in K$ and $v \in V$.
(4) $1 \cdot m = m$.   ■

Remark: Similar to the definition of *subspace*, we have the concept
of *submodule*, namely, let M be a module over R, a subset N of M
is a submodule iff (1) for any two elements $n_1, n_2 \in N$, we always
have $n_1 + n_2 \in N$ and (2) for any $n \in N$ and $r \in R$, we always have
$rn \in N$. Furthermore, we assume that the reader is familiar with the
concepts of *module homomorphism*, *module map*, *quotient*, etc.   ■

Example 2.9: Let R be any ring. Then R is a module over itself.
Any ideal I is a submodule of the R-module R. Therefore any
theorem which works for all submodules must work for all ideals.

Furthermore, we have *Nagata's*[6] *idealization principle*: let M be
an R-module. Let us define a new ring $R' = R \oplus M$ where the sum
is the usual pairwise sum, i.e., $(r, m) + (r', m') = (r + r', m + m')$
and the product is defined as $(r, m)(r', m') = (rr', rm' + r'm)$. Then
the ring R is identified with $R \oplus 0$ and the R-module $M = \{0\} \oplus M$
can be considered as an ideal of R'. Furthermore, any submodule
N of M is naturally identified with an R'-ideal $(0, N)$. Therefore
any theorem that works for all ideals must work for all submodules
N of M.

The significance of *Nagata's idealization principle* is that every
submodule N becomes an ideal of the ring $R \oplus M$ as defined above
(which is different from the usual definition of direct sum of R
and M). Together with the well-known fact that every ideal I is
a submodule of the module $R \cdot 1$, we see that the ideal theory and
submodule theory are very similar.   ■

Example 2.10: Let R be $K[x_1, \ldots, x_n]$ and let $dx_1, \ldots, dx_n$ be
symbols. Let $\Omega_{R/K} = \oplus R dx_i$. Then $\Omega_{R/K}$ is a module over R. Later
on we will see that $\Omega_{R/K}$ is a *free module* over R.   ■

---

[6]Nagata, M. Japanese mathematician. 1927–2008.

Example 2.11:

(a) Let us consider the ring $K[x,y]$ and its ideal $I = (x,y)$. It is not hard to see that x, y is a minimal generating set of I. Furthermore, the set $\{x, y\}$ is a minimal set according to inclusion in the set theory. However, they are linearly dependent, i.e.,

$$(y)x + (-x)y = 0.$$

Not as in the vector space theory, in the module theory, a minimal generating set may not be a basis.

(b) Let us consider the ideal $J = (x^2, y + xz, x + xy) \subset K[x, y, z]$. We claim that $(x^2, y + xz, x + xy) = (x, y)$. It is easy to see that $xy = x(y + xz) - z(x^2) \in J$ and $x = (x + xy) - xy \in J$ and $y = y + xz - (z)x \in J$, therefore $J = (x, y)$. We **claim** that the set $\{x^2, y + xz, x + xy\}$ is a minimal generating set of J, i.e., if we toss away any element from the set, it will not be a generating set.

(1) We toss out the first element x^2. Assume the other way. Let us consider the following equation,

$$x = f(x, y, z)(y + xz) + g(x, y, z)(x + xy).$$

where $f(x, y, z)$, $g(x, y, z)$ are polynomials in the variables. Let us replace y by -1, then $x + xy = 0$, and the above equation becomes

$$x = f(x, -1, z)(-1 + xz)$$

which implies that $(-1 + xz)|x$. It is clearly impossible.

(2) We toss out the second element $y + xz$. Assume the other way. We have

$$y = f(x, y, z)x^2 + g(x, y, z)(x + xy).$$

Let us replace x by 0. We simplify the above equation to $y = 0$ which is impossible.

(3) We toss the last element $x + xy$. Let us assume $J = (x, y) = (x^2, y + xz)$, and consider

$$x = f(x, y, z)x^2 + g(x, y, z)(y + xz).$$

Let us consider the degree 1 terms on both sides of the equation. On the left side, it is x. On the right side, it is ay for some constant a. They are clearly different. Hence it is impossible.

To summarize, we show that the set $\{x^2, y + xz, x + xy\}$ is minimal in the sense of set theory, while not minimal in the sense of cardinality. Hence it is insufficient to require the generating set to be minimal in the sense of set theory to be a *basis*, and henceforth it is impossible to deduce the *dimension* of the ideal as a module from a minimal generating set of it. ∎

Example 2.12: Let G be any commutative group. Then G is naturally a module over $\mathbb{Z}$ since for non-negative n,

$$n \cdot g = \overbrace{g + \cdots + g}^{n}$$

while for negative n, we define $n \cdot g$ as $-(-n \cdot g)$. ∎

Example 2.13: Let A be any $n \times n$ square matrix over a field K. Then we may consider A as a *linear operator* (i.e., a linear transformation from a vector space to itself) on K^n. We may define K^n as a module over $K[x]$ in the following way,

$$f(x) \cdot v = f(A) \cdot v, \quad \forall v \in K^n.$$

The K^n becomes a module over a P.I.D. $K[x]$. ∎

Example 2.14: Let A be the following 2×2 square matrix over a field $\mathbb{R}$, $A = \left(\begin{smallmatrix} 1 & 0 \\ 1 & 1 \end{smallmatrix} \right)$. Let A act on $\mathbb{R}^2$. Then we define $\mathbb{R}^2$ as a $\mathbb{R}[x]$ module by $1 \cdot v = v$ and $x \cdot [1,1]^T = A \cdot [1,1]^T = [1,2]^T$, and $[1,2]^T - [1,1]^T = [0,1]^T$, $[1,1]^T - [0,1]^T = [1,0]^T$, henceforth the module is generated by a single element $[1,1]^T$. On the other hand, let $B = \left(\begin{smallmatrix} 1 & 0 \\ 0 & 1 \end{smallmatrix} \right)$ act on $\mathbb{R}^2$. Then the module needs two generators over $\mathbb{R}[x]$. Therefore the same set $\mathbb{R}^2$ can be represented as two modules through two different matrices A, B. These two modules are not isomorphic. ∎

In the next chapter we will have further discussions of this example.

Exercises

(1) Show that every overring S of a ring R is an R-module.
(2) Show that every ideal of ring R is a submodule of R as an R-module.
(3) Let a K-vector space U be the direct sum of subspaces V and W. Show that every pair of mappings (v, w) where $v : V \mapsto T$, $w : W \mapsto T$ for a common K-vector space T. Then there is a unique map u from U to T, such that the restriction of u to V is v and the restriction of u to W is w.
(4) Let u be an operator of a vector space V. Show that if $u^2 = 0$, then $1 + u$ is an automorphism of V.
(5) Let V, W_1, W_2 be subspaces of a vector space V and V is the direct sum of V, W_1 (V, W_2 respectively). Show that W_1 and W_2 are isomorphic.
(6) Show that $\mathbb{R}^3$ can be represented as a simple $\mathbb{R}[x]$-module with $xv = Av$ where

$$
A = \begin{pmatrix} 1 & 0 & 0 \\ 1 & 1 & 0 \\ 1 & 1 & 1 \end{pmatrix}.
$$

2.3 Free Modules

As we pointed out in the preceding section that in general a minimal generating set of a module may not be linearly independent. Hence different from the theory of vector spaces, there is no basis for a general module. However, in some interesting cases of special modules we do have *basis*. We define,

Definition (Basis): A linearly independent generating set is said to be a *basis* of the module M. A free module is a module with a basis. ∎

If there is a basis $\{m_j\}_{j \in J}$ in the module M, then just as in the cases of vector spaces, the basis $\{m_j\}_{j \in J}$ produces a coordinate system: we send m_j to e_j which is an element of the standard basis of $\oplus_{j \in J} R$ with coordinate $[\ldots, 0, 0, 1, 0, 0, \ldots]$ where 1 appears at the jth place, and send any element of the form $\sum_{finite} a_j m_j$ to $[\ldots, a_j, \ldots]$. Thus M is isomorphic to $\oplus_{j \in J} R$ where R is the coefficient ring.

Proposition 2.3.1: *Let M be a free module isomorphic to $\oplus_{j \in J} R$ where R is the coefficient ring. Let I be a maximal ideal in R, and $K = R/I$ a field. Let $N = IM$. We have M/N is isomorphic to $\oplus_{j \in J} K$ with dimension the cardinality of the set J. Furthermore, any two bases of M must have the same cardinality which is called the rank of the free module M.*

Proof. It is easy to see that $N = IM$ is isomorphic to $\oplus_{j \in J} IR$, and M/N is isomorphic to $\oplus_{j \in J} K$ which is vector space over K. Therefore its dimension is the cardinality of the set J. Let $\{m'_j\}_{j \in J'}$ be another basis of M. Then any m_k can be written as $m_k = \sum_{finite} b_j m'_j$. It is easy to see that the coordinates $\{\bar{b}_j\}$ of the image of $\{m'_j\}_{j \in J'}$ in the standard basis $\{e_j\}_{j \in J}$ will generate the space $\oplus_{j \in J} K$. It is well-known that the cardinality of a generating set is bigger or equal to the dimension of a vector space, i.e.,

$$\text{cardinality}(J') \geq \text{cardinality}(J).$$

We switch the roles of J, J' and get the other inequality. Therefore we conclude that they have the same cardinality. ∎

Example 2.15: We may have a free module of infinite rank, for instance, if the ring is a field K, then an infinite dimensional vector space over K is a free module of infinite rank. ∎

Example 2.16: It is easy to see that $\mathbb{Q}$, the field of rational numbers, is not a free module over $\mathbb{Z}$, because any two non-zero elements are linearly dependent and $\mathbb{Q}$ is not generated by any singleton of one non-zero element. ∎

Example 2.17 (Complex Analysis. Double Periodic Functions): We have many functions of one period, like trigonometric functions, $\sin x$, $\cos x$, etc. Let us consider functions with double periods ω_1, ω_2, i.e.,

$$f(z + a\omega_1 + b\omega_2) = f(z), \qquad \text{for all } a, b \in \mathbb{Z}.$$

Let us consider the $\mathbb{Z}$-submodule $L = \{a\omega_1 + b\omega_2) : \text{for all } a, b \in \mathbb{Z}\}$. We call L a *lattice*. We have the following three possibilities,

(1) $\omega_1/\omega_2 \in \mathbb{Q}$. We have ω_1, ω_2 lie on the same line and they are *co-measurable*, in this case, we may reduce the two periods to one period.

(2) $\omega_1/\omega_2 \in \mathbb{R}\backslash\mathbb{Q}$. We have ω_1, ω_2 lie on the same line and they are not *co-measurable*, in other words, ω_1/ω_2 is an irrational real number. In this case, we may deduce that there are as small as possible basis for L. If we are interested only in *meromorphic* or *holomorphic* functions, then the function must be constant on the line, and hence by *analytic continuation*, it must be constant over the complex plane.

(3) The above two cases are not interesting. We will assume that $\omega_1/\omega_2 \notin \mathbb{R}$.

We are mainly interested in the above case (3). Clearly the quotient space $\mathbb{C}/L$ and $c\mathbb{C}/cL = \mathbb{C}/cL$ are isomorphic for any non-zero constant c. We may multiply ω_1, ω_2 by ω_2^{-1}, and have an equivalent pair $(\sigma, 1)$ where $\sigma = \omega_1/\omega_2$. Moreover, we may switch ω_1, ω_2, and get $(\sigma^{-1}, 1)$. Note that either $\text{im}(\sigma) > 0$ or $\text{im}(\sigma^{-1}) > 0$. We should select the order of ω_1, ω_2 so that $\text{im}(\sigma) > 0$. We conclude for every equivalent class of lattices L by non-zero c multiplications, we associate a complex number σ with $\text{im}(\sigma) > 0$ i.e., a point on the upper half-plane of $\mathbb{C}$.

Let us consider a lattice generated by (ω_1, ω_2), and the action of an element of $SL(2, \mathbb{Z})$ as follows,

$$\begin{bmatrix} a & b \\ c & d \end{bmatrix} \begin{bmatrix} \omega_1 \\ \omega_2 \end{bmatrix} = \begin{bmatrix} \omega_1' \\ \omega_2' \end{bmatrix}.$$

We have

$$\mathrm{im}(\sigma') = \mathrm{im}(\omega_1'/\omega_2') = \mathrm{im}(a\omega_1 + b\omega_2)/(c\omega_1 + d\omega_2)$$

$$= \mathrm{im}((a\omega_1 + b\omega_2)\overline{(c\omega_1 + d\omega_2)}/(c\omega_1 + d\omega_2)\overline{(c\omega_1 + d\omega_2)})$$

$$= \mathrm{im}((ad - bc)\omega_1\omega_2)/\Delta^2 > 0.$$

The last part is due to the fact that $ad - bc = 1$ and $\mathrm{im}(\omega_1/\omega_2) > 0 \iff \mathrm{im}(\omega_1\omega_2) > 0$. ∎

Example 2.18 (Guo, Shou-jing[7]): A constant polynomial may be viewed as an Arithmetic series of grade 0. A linear polynomial $f(x) = ax + b$ may be viewed as an Arithmetic series (of grade 1), when we take the value of x at successive positive integers. If we take the difference $f(n) - f(n - 1) = a$, then we get constant a which is considered to be an Arithmetic series of grade 0. In general, we defined an Arithmetic series $f(n)$ of grade m by the difference $f(n) - f(n-1)$ being an Arithmetic series of grade $m-1$. It is obvious that an Arithmetic series of grade m is a polynomial of degree m.

In Guo's work about astronomy to establish the *Shoushi Calendar*, given the data he had to find the fitting polynomial, in other words, he had to find a polynomial $f(x)$ knowing its degree d and values at $d+1$ equal distance consecutive points. Let us take one example. Let the degree d be ≤ 3, and its values at $0, 1, 2, 3$ be $0, 5, 12, 3$, then we have the following table,

$$\begin{bmatrix} a_i : & 0, & 5, & 12, & 3 \\ b_i = a_i - a_{i-1} : & 5, & 7, & -9 & \\ c_i = b_i - b_{i-1} : & 2, & -16 & & \\ d_1 = c_i - c_{i-1} : & -18 & & & \end{bmatrix}.$$

Then the polynomial $f(x) = d_0 C_3^x + c_0 C_2^x + b_0 C_1^x + a_0 = -18 C_3^x + 2C_2^x + 5C_1^x + 0C_0^x$ where

$$C_n^x = \frac{x(x-1)\cdots(x-n+1)}{n(n-1)\cdots 1}.$$

---

[7]Guo, Shou-jing. Chinese astronomer, engineer and mathematician. 1231–1316. (The asteroid 2012 was named after him.) He further developed the arithmetical theory of polynomials started by Shen, Kua.

Definition We define the *Guo ring* as the ring of all polyno-mials $\in \mathbb{Q}[x]$ which have integer values at integer points (see the Exercises). ∎

Guo can be viewed as the pioneer of the *theory of finite differences* and *Hilbert's*[8] *polynomials*. Its importance is obvious.

We study the finite rank cases, and the following proposition is interesting,

Proposition 2.3.2: *Let $A \in M_{n \times n}(R)$ be the square matrices with entries in R. Then A is invertible iff* $\det(A)$ *is a unit in R.*

Proof. ($\Leftarrow$) If $\det(A)$ is a unit in R, then by the usual cofactor, adjoint formula, we can easily find A^{-1}.

($\Rightarrow$) If the inverse of A exists, call it B, then we have

$$AB = I.$$

Therefore we have

$$\det(A) \det(B) = 1$$

so $\det(A)$ must be a unit in R. ∎

Let M be a free module of rank n with a basis $\{m_1, \ldots, m_n\}$. It is easy to see that another set $\{n_1, \ldots, n_n\}$ is a basis of M iff the relation matrix A from $\{m_1, \ldots, m_n\}$ to $\{n_1, \ldots, n_n\}$ is invertible.

Example 2.19: Let G be any commutative group. Then G is naturally a module over $\mathbb{Z}$ since for non-negative n,

$$n \cdot g = \overbrace{g + \cdots + g}^{n}$$

while for negative n, we define $n \cdot g$ as $-(-n \cdot g)$. ∎

Exercises

(1) Show that the Guo ring is a module over $\mathbb{Z}[x]$.
(2) Show that $K[x, y]$ is not a P.I.D.

---

[8]Hilbert, D. German mathematician. 1862–1943.

(3) Show that a group of order 15 must be commutative.

(4) A ring is said to be *noetherian* if every ideal is finitely generated. Show that a ring is noetherian iff any submodule of any finitely generated module over it is finitely generated.

(5) In the relation $\mathbb{Z}[x] \subset$ Guo ring $\subset \mathbb{Q}[x]$, $\mathbb{Z}[x]$, $\mathbb{Q}[x]$ are noetherian and the Guo ring is not. For definition of the noetherian ring, see the preceding Exercises.

(6) (*Nagata's idealization principle*): We have to be careful in applying the principle. Show that if R is a P.I.D. and M an R-module, the newly-defined ring $R' = R \oplus M$ in the description of Nagata's idealization principle may not be a P.I.D.

(7) (*Nagata's idealization principle*): Show that if R is a noetherian ring and M a finitely generated R-module, the newly-defined ring $R' = R \oplus M$ is a noetherian ring.

(8) Show that every commutative group is a $\mathbb{Z}$-module.

2.4 Finitely Generated Modules over a P.I.D.

For our convenience, let us define,

Definition If there is a finite generating set of an R-module M, then M is said to be a *finitely generated module* or a *finite module*. ∎

Let R be a P.I.D. and L be a finite R-module generated by a set $\{\ell_1, \ldots, \ell_n\}$. Let us define a free R-module $M = \oplus_1^n R e_i$ where e_i's are symbols. We have a map $\phi : M \mapsto L$ as

$$\phi(e_i) = \ell_i.$$

It is easy to verify that ϕ is onto. Let its kernel be N.

As usual, $L \approx M/N$. To understand L is nothing, but rather to understand M, N and the relation between M and N. Before going further, we want to generalize the class of P.I.D. to the class of *noetherian*[9] *rings*, which are defined as

---

[9]Noether, E. German female mathematician and physicist. 1882–1935.

Definition (Noetherian Ring): Given a ring R. The following three conditions are equivalent, where each of them defines a *noetherian ring*,

(1) Every ideal of R is finitely generated.
(2) The *ascending chain condition* is satisfied, i.e., every ascending chain of ideals $I_0 \subset I_1 \subset \cdots \subset I_i \subset \cdots$ must terminate, that is to say there is a positive integer n such that $I_n = I_{n+1} = \cdots = I_m = \cdots$.
(3) *Maximal principle*: In every non-empty family $\mathfrak{S}$ of ideals, there must be a maximal one. ∎

Remark: We may stick to P.I.D. with all mentioned *noetherian rings* below to be replaced by *P.I.D.* However the new proofs are not simpler by this replacement. The definition of noetherian ring is common. The reader is referred to any book on algebra about the equivalence of the above three conditions. ∎

We have the following proposition,

Proposition 2.4.1: *Let R be a noetherian ring, and M a finite R-module. Then any submodule N of M is a finite module.*

Proof. Let $M = \sum_{i=1}^{q} Rm_i$. Let $I = \{a_1 : a_1 m_1 + \sum a_i m_i \in N$, for some $a_i, i > 1\}$. Then clearly I is an ideal in R. Therefore I is finitely generated, say $I = (a_{11}, a_{12}, \ldots, a_{1s})$, and $n_1 = a_{11} m_1 + \cdots \in N, n_2 = a_{12} m_1 + \cdots \in N, \ldots, n_s = a_{1s} m_1 + \cdots \in N$. Let $N_1 = \sum_{i>1} Rm_i$. It is clear that for any $n \in N$ with $n = am_1 + \cdots$, then we have $a \in I$, and we can find f_i such that $a = \sum f_i a_{1i}$. Then it is easy to see that $n - \sum_i f_i n_i \in N_1$. Now by making a reduction on the number $q \geq 1$, our proposition is proved. ∎

For a Euclidean domain, say $\mathbb{Z}$ or $K[x]$, given two elements a, b, the *long algorithm* based on the Euclidean operations suffices to reach the GCD of a, b by repeatedly subtracting a multiple of one of a, b from the other one. However, for a general P.I.D. (which is not Euclidean), the corresponding Gaussian process is not enough.

We need the following general operations. Note that the Gaussian operations can be generalized as follows,

Definition Given any $m \times m$ matrix A, we define an *elementary invertible matrix* as an invertible matrix C of the following forms,

(1) In the identity matrix I we replace one of the 1's on the diagonal by a unit a in R, as

$$C = \begin{bmatrix} 1 & 0 & \cdot & \cdot & \cdot & \cdot & \cdot & 0 \\ 0 & 1 & \cdot & \cdot & \cdot & \cdot & \cdot & \cdot \\ \cdot & \cdot & \cdot & \cdot & \cdot & \cdot & \cdot & \cdot \\ 0 & 0 & \cdot & a & \cdot & 0 & \cdot & \cdot \\ \cdot & \cdot & \cdot & \cdot & \cdot & \cdot & \cdot & \cdot \\ 0 & 0 & \cdot & 0 & \cdot & 1 & \cdot & 0 \\ 0 & 0 & \cdot & \cdot & \cdot & \cdot & 1 & 0 \\ 0 & 0 & \cdot & \cdot & \cdot & \cdot & 0 & 1 \end{bmatrix}.$$

(2) In the following matrix C, $ad - bc = \epsilon$ with ϵ a unit in R,

$$C = \begin{bmatrix} 1 & 0 & \cdot & \cdot & \cdot & \cdot & \cdot & 0 \\ 0 & 1 & \cdot & \cdot & \cdot & \cdot & \cdot & \cdot \\ \cdot & \cdot & \cdot & \cdot & \cdot & \cdot & \cdot & \cdot \\ 0 & 0 & \cdot & a & \cdot & b & \cdot & \cdot \\ \cdot & \cdot & \cdot & \cdot & \cdot & \cdot & \cdot & \cdot \\ 0 & 0 & \cdot & c & \cdot & d & \cdot & 0 \\ 0 & 0 & \cdot & \cdot & \cdot & \cdot & 1 & 0 \\ 0 & 0 & \cdot & \cdot & \cdot & \cdot & 0 & 1 \end{bmatrix}$$

where the number a is on the (ii) position, the number b is on the (ij) position, the number c is on the (ji) position, the number d is on the (jj) position, all entries on positions (kk) are 1 if $k \neq i, j$, all other entries are 0. ∎

Proposition 2.4.2: *The inverse of an elementary invertible matrix is an elementary invertible matrix.*

Proof. Trivial. ∎

Remark: It is easy to see that the elementary matrices in terms of row or column operations are elementary invertible matrices. What we will state and prove in the *Smith Normal Form* is a generalization of the *Normal Form* of matrices. ∎

Lemma 2.4.3: *Given any two elements* $\alpha, \beta \in$ *P.I.D. R with* $(\alpha, \beta) = (\gamma)$. *Then we may use an elementary invertible matrix multiplying from the left to send* $[\alpha, \beta]^T \mapsto [\gamma, 0]^T$.

Proof. For any two elements α, β, we find their gcd as

$$\alpha a + \beta b = \gamma$$

$$\alpha = d\gamma, \quad \beta = -c\gamma$$

$$ad - bc = 1.$$

We shall construct the mapping by the matrix $\begin{bmatrix} a & b \\ c & d \end{bmatrix}$. The mapping φ represented by the above matrix is the one required by the lemma. ∎

Let R be a P.I.D. In the next theorem we deduce some useful form for any matrix $A = (a_{ij}) \in M_{n \times m}(R)$. Note that we conclude in the theorem that the ideal generated by the left corner entry a'_{11} be uniquely determined by the original matrix A, for the uniqueness of the remaining terms $a'_{22}, \ldots, a'_{\ell\ell}$; please see the next section.

The following theorem is in honor of Henry J. S. Smith, an Irish mathematician of the 19th century,

Theorem 2.4.4 (Smith Normal Form): *Let* $A = (a_{ij}) \in M_{n \times m}(R)$ *where R is a P.I.D. and $A \neq (0)$. Let ideal* $(a_{11}, \ldots, a_{nm}) = (a)$. *Then we may use the invertible maps to change A to a matrix $A' = (a'_{ij})$, i.e., there are invertible matrices C, D with* $CAD = A' = (a'_{ij})$ *such that*

(1) $a'_{ij} = 0$ *if* $i \neq j$.
(2) $a = a'_{11}$.
(3) $a'_{11} | a'_{22} | \cdots | a'_{\ell\ell}$, $a'_{jj} \neq 0$, $\forall j \leq \ell$; $a'_{jj} = 0$, $\forall j > \ell$.

Proof. Note that the operations we apply are all invertible, therefore the ideal $(a_{11}, \ldots, a_{nm})$ does not change, so all entries in any stage of the following induction process generate (a). The above theorem is clearly true for $n = 1$ or $m = 1$. We wish to use mathematical induction for $n, m > 1$. Let us consider the first row, take any two elements a_{1i}, a_{1j}, and we find their gcd as

$$aa_{1i} + ba_{1j} = \gamma$$

$$a_{1i} = d\gamma, \quad a_{1j} = -c\gamma$$

or

$$ad - bc = 1.$$

We may apply the following elementary invertible matrix multiplied from the right, $\begin{bmatrix} a & c \\ b & d \end{bmatrix}$ to the ith column and jth column. It is not hard to see that after finitely many steps, we make the first row $[b_1, 0, \ldots, 0]$, then we treat the first column in a similar way, certainly we shall multiply matrices from the left, and switch the position of b, c. Two things may happen; (1) b_1 is a factor of the particular coefficient we are treating, then we create a 0 at the spot without disturbing the 0's we created for the first row, or (2) b_1 is not a factor of the particular coefficient we are treating, then we have to use an elementary invertible matrix to create a new leading element b_2 of the first row such that (b_1) is a proper sub-ideal of (b_2). Now we repeat the process of clearing the first row, with the new leading coefficient b_3. Then we have $(b_3) \supset (b_2)$. Therefore (b_1) is a proper sub-ideal of (b_3). Since R is a P.I.D., hence a noetherian ring, if we can repeat the above process (2), then we construct an ascending chain of ideals which must terminate. Therefore we reach a situation that the matrix is modified to the following form

$$A' = \begin{bmatrix} b & 0 & \cdot & \cdot & \cdot & 0 \\ 0 & \bar{a}_{22} & \cdot & \cdot & \cdot & \cdot \\ \cdot & \cdot & \cdot & \cdot & \cdot & \cdot \\ 0 & \cdot & \cdot & \cdot & \cdot & \cdot \\ 0 & \cdot & \cdot & \cdot & \cdot & \cdot \end{bmatrix}.$$

By mathematical induction on the integer n, m, we may assume that the smaller matrix of $(n-1) \times (m-1)$ can be further modified to the following matrix,

$$A'' = \begin{bmatrix} b & 0 & \cdot & \cdot & 0 & 0 \\ 0 & \bar{a}_{22} & \cdot & \cdot & 0 & 0 \\ \cdot & \cdot & \cdot & \cdot & 0 & \cdot \\ 0 & \cdot & \cdot & \bar{a}_{\ell\ell} & 0 & 0 \\ 0 & 0 & \cdot & 0 & 0 & 0 \end{bmatrix}$$

satisfies $\bar{a}_{22}|\bar{a}_{33}| \cdots |\bar{a}_{\ell\ell}$. If $b|\bar{a}_{22}$, then we are done. If not, we add the second row to the first row, and then apply a suitable elementary invertible matrix to put the GCD$(b, \bar{a}_{22})$ on the first column and 0 on the second column. Note that the second entry of the first column is not 0 anymore. However it can be easily cleaned up using the new first row. After we find the final form of the matrix, we shall name them $a'_{11}, a'_{22}, \ldots, a'_{\ell\ell}$. Thus we reach the form required by the Smith Normal Form. The only thing that remains to be shown is that the new a'_{11} is a generator of the ideal of all entries. For this purpose, we simply point out that the ideal is invariant under all invertible matrix multiplications, and it is clear at the end we have the ideal of all entries that is generated by a'_{11}, thus we conclude that the ideal generated by all entries at the very beginning is generated by the new a'_{11}. ∎

Exercises

(1) Prove Proposition 2.4.2.
(2) Show that $\mathbb{Z}[x]$ is not a P.I.D.
(3) Let $f(x) = x$, $g(x) = 1 + x^2$. Find a 2×2 matrix A such that

$$A \cdot \begin{bmatrix} f(x) \\ g(x) \end{bmatrix} = \begin{bmatrix} 1 \\ 0 \end{bmatrix}.$$

(4) Find the Smith Normal Form of the following matrix A with coefficients in $\mathbb{Q}$,

$$A = \begin{bmatrix} 5 & 4 & 2 & 8 \\ 6 & 2 & 2 & 2 \\ 0 & 8 & 4 & 6 \end{bmatrix}.$$

(5) Show that an R-module M is finitely generated if and only if there is a surjective map $\rho : R^n \mapsto M$ for some n.

(6) Find the Smith Normal Form of the following matrix A with coefficients in $\mathbb{R}[x]$,

$$A = \begin{bmatrix} x & 1+x & 1+x^2 & 1+x^3 \\ 6 & 1 & 1+x & 1+x^2 \\ 0 & 0 & 1 & 1+x \end{bmatrix}.$$

2.5   An Application of Smith Normal Form to Coding Theory

Convolutional Codes

Representation

The concept of convolution code was introduced by P. Elias[10] (*Coding for Noisy Channels*, IRE Conv. Record, Part 4, pp. 37–46, Einhoven University of Technology, 1993). Any sequence of data or a stream of data $(a_0, a_1, \ldots, a_n, \ldots)$ with $a_i \in \mathbf{F}$ can be represented as a power series,

$$f(x) = \sum_{0}^{\infty} a_i x^i.$$

Let us consider the coding process over $\mathbf{F}_2$. Given a stream over $\mathbf{F}_2$, $(a_0, a_1, a_2, a_3, \ldots, a_n, \ldots)$, i.e., $a_i \in \mathbf{F}_2$, we may use the following process to produce two (in general, maybe one, or any finitely many) data streams $(b_0, b_1, \ldots, b_n, \ldots)$, $(c_0, c_1, \ldots, c_n, \ldots)$.

---

[10]Elias, P. American information theorist. 1923–2001.

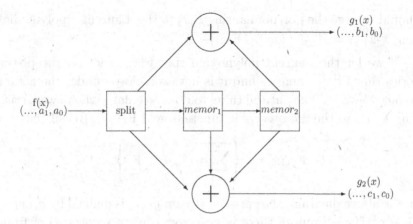

Where at the time 0, a_0 enters the leftmost square at the middle and the other two squares at the middle row are with 0's, the two $\oplus$ at the top and bottom denote module 2 additions. As time goes from n to $n+1$, all values in the middle boxes move to their right, so the rightmost one fall off and disappear, the $(n+1)$th coefficient will occupy the leftmost box.

Mathematically, let us use

$$g_1(x) = \sum_0^\infty b_i x^i, \qquad g_2(x) = \sum_0^\infty c_i x^i$$

where

$b_0 = a_0,\ b_1 = a_1,\ b_n = a_n + a_{n-2}$ for all $n > 1$

$c_0 = a_0,\ c_1 = a_1 + a_0,\ b_n = a_n + a_{n-1} + a_{n-2}$ for all $n > 1$.

Or we simply write

$$g_1(x) = (1 + x^2)f(x)$$
$$g_2(x) = (1 + x + x^2)f(x).$$

It seems that an encoder is simply a multiplication by a polynomial. However there is a catch: a multiplication by x (a delay of time) should be considered as invertible (at least left-invertible)! So we should enlarge the polynomial ring $\mathbf{F}[x]$ to the Laurent[11] polynomial ring $\mathbf{F}[x]_{(x)} = \{\frac{h(x)}{x^d} : h(x) \in \mathbf{F}[x]\}$.

If we let the Laurent polynomial ring $\mathbf{F}[x]_{(x)}$ act on the power series ring $\mathbf{F}[[x]]$, then we find it is not even closed under the action induced by x^{-1}. The natural thing to do is to enlarge the power series ring $\mathbf{F}[[x]]$ to the meromorphic function field $\mathbf{F}((x))$. Recall that

$$\mathbf{F}_2((x)) = \left\{ \sum_{-m}^{\infty} a_i x^i : a_i \in \mathbf{F}_2 \right\}.$$

In the above diagram, the encoded stream $g_1(x)$ is defined by $g_1(x) = (1 + x^2)f(x)$. Suppose there is no error, can we recover $f(x)$ from $g_1(x)$? Mathematically, $f(x) = (1 + x^2)^{-1}g_1(x)$, however, there are two problems: (1) The recursive formula $a_0 = b_0, a_1 = b_1, a_2 = b_2 - b_1,$ $a_3 = b_3 - b_2 + b_1, \ldots$ is getting longer and longer without a bound. Therefore, we require an unbounded number of memory units. (2) If we have a single error for $g_1(x)$, namely replacing $g_1(x)$ by $g_1(x) + x^n$, then the inverse will differ with $f(x)$ at infinitely many places. Any encoder with the second type problem will be called a *catastrophic encoder* and will not be used.

Combining and Splitting

It is clear that the only good encoders which take one incoming stream of data $f(x)$ and produce one stream of data $g_1(x)$ are multiplying by x^n, i.e., $g(x) = x^n f(x)$. Those are non-interesting. How about an encoder which produces more streams of data as in our preceding diagram? Let us first study the technique of combining and splitting data streams. Let

$$f(x) = \sum_{i=-m}^{\infty} a_i x^i$$

$$g_j(x) = \sum_{i=-m}^{\infty} b_{ji} x^i \quad \text{for } j = 1, 2, \ldots, r.$$

---

[11]Laurent, P. French female mathematician. 1813–1854.

Then we have $h_j(x), h(x)$ uniquely defined by the following equations,

$$f(x) = \sum_{j=0}^{n-1} x^j h_j(x^n)$$

$$h(x) = \sum_{j=1}^{r} x^j g_j(x^r).$$

In other words, we may split one stream of data $f(x)$ into n streams of data $h_j(x)$ for $j = 1, 2, \ldots, n$, in symbol $S_n(f(x)) = [h_0(x), h_1(x), \ldots, h_{n-1}(x)]$, and combining r streams of data $g_j(x)$ for $j = 1, 2, \ldots, r$ into one stream of data $h(x)$, in symbol $C_r(g_1(x), \ldots, g_r(x)) = h(x)$. The splitting and combining operations are one-to-one and onto maps between $\mathbf{F}((x))$ and $\mathbf{F}((x))^n$ or $\mathbf{F}((x))^r$ while they are non-linear with respect to the field $\mathbf{F}((x))$.

Smith Normal Form

Our preceding figure over $\mathbf{F}_2$ in the subsection of representation of convolution codes can be written mathematically as

$$[f(x)][1 + x^2, 1 + x + x^2] = [g_1(x), g_2(x)].$$

Using some linear algebra, we may rewrite the above matrix equations as

$$[f(x)] \cdot [1] \cdot [1 \quad 0] \cdot \begin{bmatrix} 1 + x^2 & 1 + x + x^2 \\ x & 1 + x \end{bmatrix} = [g_1(x) \quad g_2(x)].$$

In other words, we have the following matrix equation

$$[1 + x^2 \quad 1 + x + x^2] = [1] \cdot [1 \quad 0] \cdot \begin{bmatrix} 1 + x^2 & 1 + x + x^2 \\ x & 1 + x \end{bmatrix}.$$

The above is the well-known *Smith Normal Form* of a matrix over a P.I.D. $\mathbf{F}_2[\mathbf{x}]$ follows as,

Proposition 2.5.1: *Let R be a P.I.D., and M an $r \times n$ matrix with entries in R. Then M can be written as*

$$M = A\Gamma B$$

where (1) *both A and B are invertible such that their inverses are with entries in R.* (2) Γ *is in the diagonal form with entries on the diagonal the invariant factors γ_i of M. Note that $\gamma_1|\gamma_2|\cdots|\gamma_r$.*

Proof. Omitted.   ∎

In our present example, we have the ground field F_2 and $A^{-1} = [1]$ and

$$B^{-1} = \begin{bmatrix} 1+x^2 & 1+x+x^2 \\ x & 1+x \end{bmatrix}.$$

Therefore we have the following identity,

$$[g_1(x) \quad g_2(x)] \cdot \begin{bmatrix} 1+x^2 & 1+x+x^2 \\ x & 1+x \end{bmatrix} \cdot \begin{bmatrix} 1 \\ 0 \end{bmatrix} \cdot [1] = [f(x)].$$

We conclude that with only finitely many memory units (in fact, at most 4), we may recover $f(x)$ if there is no error for $g_1(x)$ and $g_2(x)$. Furthermore, if there are single errors for $g_1(x)$ and $g_2(x)$, say replacing them by x^n, x^m, then the decoding result is a polynomial which is not an infinitely long meromorphic function. Therefore the encoder is not a catastrophic encoder.

In general, we may consider r streams of data $\mathbf{J} = [f_1(x), \ldots, f_r(x)]$ where $f_i(x) \in \mathbf{F}((x))$, we may view it as a vector J in the r-dimensional vector space $\mathbf{F}((x))^r$. An encoder may be viewed as an $r \times n$ matrix M with coefficients in the Laurent polynomial ring $\mathbf{F}[x]_{(x)}$ to produce n streams of data as a vector T in $\mathbf{F}((x))^n$ in the following formula,

$$JM = T.$$

According to the above Smith Normal Form, the above equation can be rewritten as

$$JA\Gamma B = T.$$

Since both A, B are invertible, Γ is uniquely determined by M. If Γ is right invertible $\Longleftrightarrow$ invariant factors of M are invertible, i.e., they are powers of x. Then and only then we have a non-catastrophic encoder.

Exercises

(1) Is the convolution code over F_2 defined by $[f(x)][1+x^2, 1+x+x^2]$ catastrophic?
(2) Is the convolution code over F_2 defined by $[f(x)][1+x, 1+x+x^2+x^3]$ catastrophic?

2.6 Fundamental Theorem of Finitely Generated Modules over a P.I.D.

A field is a P.I.D. We apply the Smith Normal Form in the case of a field. In this case we get the *normal form* of any matrix A.

Example 2.20: Let us consider the following matrix with integer coefficients,

$$A = \begin{bmatrix} 3 & 2 & 5 \\ 4 & 2 & 8 \\ 0 & 6 & 10 \end{bmatrix}.$$

It is easy to see that $(a_{11}, \ldots, a_{33}) = (1)$. Let us use the Gaussian column operation $c_1' = c_1 - c_2$, then we use $a_{11}' = 1$ to clear all entries of the first row and first column. Then we get

$$A' = \begin{bmatrix} 1 & 0 & 0 \\ 0 & -2 & -2 \\ 0 & 18 & 40 \end{bmatrix}.$$

Now we use the Gaussian column operation replacing c_3' by $c_3' - c_2'$ and using row operation to clear the term with position $(3, 2)$, and we have

$$A' = \begin{bmatrix} 1 & 0 & 0 \\ 0 & -2 & 0 \\ 0 & 0 & 22 \end{bmatrix}.$$

We get the Smith Normal Form of A. ■

Example 2.21: Let us consider the following matrix A over $R = \{a + b(1 + \sqrt{-19})/2 : a, b \in \mathbb{Z}\}$ which is a non-Euclidean P.I.D.

(cf. T. T. Moh, *Algebra*, p. 151). It is easy to verify that the set of units $= \{1, -1\}$. Let $\alpha = 2$, $\beta = 1 + \sqrt{-19}/2$, $(\alpha, \beta) = (a)$ and let $A = [\alpha, \beta]$. Then we may argue that no sequence of the Gaussian column operations will reduce the matrix A to the form $[a, 0]$.

It is easy to see that our new elementary invertible matrices are required.   ∎

Let us use the following notations,

Notation: To avoid confusion, let M be a finite free module over a noetherian ring R with a basis $\{e_1, \ldots, e_m\}$. Let $\mathbf{e} = [e_1, \ldots, e_m]^T \in \bigoplus_1^m M$. Let N be a finite submodule of M with generating set $\{n_1, \ldots, n_n\}$. We may write $\mathbf{n} = [n_1, n_2, \ldots, n_n]^T$. Let $n_j = \sum_{i=1}^m a_{ji} e_i$. We shall write $[n_1, \ldots, n_n]^T$ as $(a_{ji})\mathbf{e}$ where (a_{ji}) is an $n \times m$ matrix A. We may write $\mathbf{n} = A\mathbf{e}$.   ∎

Proposition 2.6.1: *Let us use the above notation. Let $\{e_1, \ldots, e_m\}$ be a basis for M and $\{n_1, \ldots, n_n\}$ be a generating set for N. Let C be an $m \times m$ invertible matrix, and D an $n \times n$ invertible matrix. Let $\mathbf{e}' = C^{-1}\mathbf{e}$, and $\mathbf{n}' = D\mathbf{n}$. Then $\mathbf{e}'$ is a basis for M, and $\mathbf{n}' = D\mathbf{n}$ is a generating set for N. Furthermore, assume $\mathbf{n} = A\mathbf{e}$, then $\mathbf{n}' = DAC\mathbf{e}'$.*

Proof. It is easy to see that $D\mathbf{n}$ and $C^{-1}\mathbf{e}$ are generating sets for M and N, respectively. Let us show that $C^{-1}\mathbf{e} = \mathbf{e}'$ is a linearly independent set. If $\mathbf{e}'$ is not a linearly independent set, let the following,

$$\sum a_i e_i' = 0$$

be a linear equation. Then it can be rewritten as

$$[a_1, \ldots, a_m] C^{-1}\mathbf{e} = 0.$$

Since $\mathbf{e}$ is a linearly independent set, we must have

$$[a_1, \ldots, a_m] C^{-1} = 0.$$

Furthermore C^{-1} is an invertible matrix, we must have

$$[a_1, \ldots, a_m] = [0, \ldots, 0].$$

In other words, all $a_i = 0$, and $\mathbf{e}'$ is a linearly independent set.   ∎

Theorem 2.6.2: *Let R be a P.I.D. and M a finitely generated free module of rank n over R, and N a non-zero submodule of M. Then there is a basis $\{m_1, m_2, \ldots, m_n\}$ of M and $c_1, c_2, \ldots, c_\ell \in R$ such that,*

(1) $c_1|c_2| \cdots |c_\ell$ *and* $\ell \leq n$.
(2) N *is generated by* $c_1m_1, c_2m_2, \ldots, c_\ell m_\ell$.

Proof. Let us make an induction argument on the rank of M. If $\text{rank}(M) = 1$, then the theorem is obvious. Let us consider the case $\text{rank}(M) = n \geq 2$. Take any basis $\{\bar{m}_i\}$. Let us look at the coordinates $\{[a_{i1}, \ldots, a_{ij}, \ldots, a_{in}]^T\}$ of any finite generating set $\{n_i : i = 1, \ldots, m\}$ of N. We thus form a matrix $A = (a_{ij})$ whose column vectors are those coordinate vectors. Note that $n_i = \sum a_{ij}\bar{m}_j$, and $[n_1, \ldots, n_m]^T = A \cdot \mathbf{m}^T$ where $\mathbf{m} = \{\bar{m}_1, \ldots, \bar{m}_n\}$. By the theorem of the Smith Normal Forms, there are invertible matrices C, D such that $DAC = A' = (a'_{ij})$ is of the form specified by the Smith Normal Form. The operation of multiplying from the left by D can be understood as a changing of the generating set, and the multiplying by C on the right is understood as $DAm = A'C^{-1}\mathbf{m} = A'\mathbf{m}'$, i.e., a changing of the basis of M without changing the generating set of N. Therefore the Smith Normal Form can be understood as changing the basis of M (without changing the generating set of N) and changing the generating set of N (without changing the basis of M). Our theorem is thus proved. ∎

A finite free module over a P.I.D. resembles a finite dimensional vector space over a field. In fact we have,

Corollary 2.6.3: *If M is a finite free module over a P.I.D. R, then any submodule N of M is free and with* $\text{rank}(N) \leq \text{rank}(M)$.

Proof. It follows from the preceding theorem trivially. ∎

Corollary 2.6.4: *The ideal (c_1) is uniquely determined by N.*

Proof. It follows from the preceding theorem trivially. ∎

Example 2.22: Let us consider the submodule N of $\mathbb{Z}^3$ generated by $[3, 4, 0]^T$, $[2, 2, 6]^T$, $[5, 8, 10]^T$, $[2, 6, -22]^T$. According to our theorem we may consider the following matrix with integer coefficients,

$$A = \begin{bmatrix} 3 & 2 & 5 & 2 \\ 4 & 2 & 8 & 6 \\ 0 & 6 & 10 & -22 \end{bmatrix}.$$

It is easy to see that $(a_{11}, \ldots, a_{34}) = (1)$. Let us use the Gaussian column operation $c_1' = c_1 - c_2$, then we use $a_{11}' = 1$ to clean all entries of the first row and first column, and we get

$$A' = \begin{bmatrix} 1 & 0 & 0 & 0 \\ 0 & -2 & -2 & 2 \\ 0 & 18 & 40 & -16 \end{bmatrix}.$$

Now we use the Gaussian column operations $c_3' - c_2'$ and $c_4' + c_2'$, then clean the second column of the smaller matrix, and we have

$$A' = \begin{bmatrix} 1 & 0 & 0 & 0 \\ 0 & -2 & 0 & 0 \\ 0 & 0 & 22 & 2 \end{bmatrix}.$$

We get the Smith Normal Form of A,

$$A' = \begin{bmatrix} 1 & 0 & 0 & 0 \\ 0 & -2 & 0 & 0 \\ 0 & 0 & 2 & 0 \end{bmatrix}.$$

So we conclude that under an automorphism ϕ of $\mathbb{Z}^3$, $\phi(N)$ is generated by three elements $[1, 0, 0]^T$, $[0, -2, 0]^T$, $[0, 0, 2]^T$. ∎

Theorem 2.6.5 (Fundamental Theorem of Finitely Generated Modules over a P.I.D.): *Let R be a P.I.D. and M a finitely generated module over R, there exist non-zero elements $c_1, c_2, \ldots, c_\ell \in R$ such that*

(1) $c_1 | c_2 | \cdots | c_\ell$.

(2) *M is isomorphic to $R/(c_1) \oplus R/(c_2) \oplus \cdots \oplus R/(c_\ell) \oplus R \oplus \cdots \oplus R$.*

Proof. Let M be generated by $g_1, \ldots, g_n$. Then M can be represented by

$$\text{Ker}(\phi) \xrightarrow{\quad i \quad} \oplus_j \text{Re}_j \xrightarrow{\quad \phi \quad} M$$

where $\oplus_j \text{Re}_j$ is a free module generated by $\{e_j\}$, $\phi : e_j \mapsto g_j$, i is the natural embedding map which sends $\text{Ker}(\phi)$ to $\oplus_j \text{Re}_j$. Naturally, we have $M \cong \oplus_j \text{Re}_j / \text{Ker}(\phi)$. According to the previous theorem, we may select a basis $\{m_1, \ldots, m_n\}$ for $\oplus_j \text{Re}_j$ such that there are elements $c_1 | c_2 | \cdots | c_\ell$ such that $\text{Ker}(\phi)$ is generated by $c_1 m_1, \ldots, c_\ell m_\ell$. Therefore we have

$$M \approx R/(c_1) \oplus \cdots \oplus R/(c_\ell) \oplus R \oplus \cdots \oplus R. \qquad \blacksquare$$

Note that if $c_i = 1$, then $(c_i) = R$ and $R/(c_i) = 0$, usually we omit the first few $c_i = 1$. Thus we assume $\mathbf{c_1}$ is a non-unit. In the case that R is a field, hence all c_i are units, we just have

$$M \approx R \oplus \cdots \oplus R$$

which is a finite dimensional vector space over R.

Corollary 2.6.6: *The ideal (c_1) is uniquely determined by M.*

Proof. It follows from the preceding theorem trivially. $\qquad \blacksquare$

Definition The factor $c_1, c_2, \ldots, c_\ell$ are called the *torsion factors* of the finitely generated module M. Later on in the next section, we will establish the uniqueness of them. The submodule of M which maps to $R/(c_1) \oplus \cdots \oplus R/(c_\ell)$ is called the *torsion submodule*, $\text{Tor}(M)$, of M. It can be defined as

$$\{n \in M : \exists\, 0 \neq r \in R \ni: rn = 0\}$$

as long as the ring R is a domain. The number of R in $R \oplus \cdots \oplus R$ in the preceding theorem is called the *Betti number* of M, or the Betti number $= \text{rank}(M/(\text{torsion submodule of } M))$. $\qquad \blacksquare$

Corollary (of the Theorem): *Let the K-vector space V be n-dimensional. Given an $n \times n$ matrix A. Let $R = K[x]$ and V*

is represented as R-module through the action of A, i.e., we define $xv = Av$. Then V is isomorphic to $R/(c_1) \oplus R/(c_2) \oplus \cdots \oplus R/(c_\ell)$, in other words, in the fundamental theorem, the Betti number must be zero.

Proof. It follows from the preceding theorem trivially, since the dimension of Kx over K is infinite, then $\oplus K[x]$ cannot be a subspace of a finite dimensional vector space over K. ∎

Proposition 2.6.7: *The Betti number is invariant for a finitely generated module M over a P.I.D. R.*

Proof. Clearly the Betti number of M = the rank of $M/\mathrm{Tor}(M)$ which is the rank of the free module $M/\mathrm{Tor}(M)$, which is invariant. ∎

Example 2.23: Let an abelian group $Z/(12) \oplus Z/(18) \oplus Z$ be given. Then according to the preceding theorem it is isomorphic to $Z/(6) \oplus Z/(36) \oplus Z$ with torsion factors $6, 36$ and Betti number $= 1$. ∎

Exercises

(1) Show that every Euclidean domain is a P.I.D.
(2) Find the Smith Normal Form of the following matrix A with coefficients in $\mathbb{Z}$,

$$A = \begin{bmatrix} 5 & 4 & 3 & 8 \\ 1 & 2 & 2 & 2 \\ 1 & 8 & 4 & 6 \end{bmatrix}.$$

(3) Find the Smith Normal Form of the following matrix A with coefficients in $\mathbb{R}[x]$,

$$A = \begin{bmatrix} x^2 & x^2 + x & x + x^2 & x + x^3 \\ x & 1 & x^4 + x & x^3 + x^2 \\ x & x & x^2 & x^4 + x \end{bmatrix}.$$

(4) Let U be the submodule of $\mathbb{R}[x] \oplus \mathbb{R}[x] \oplus \mathbb{R}[x]$, generated by $[x^2, x, x]^T, [x^2 + x, 1, x]^T, [x + x^2, x^4 + x, x^2]^T, [x + x^3, x^3 + x^2, x^4 + x]^T$. Simplify the system of generators.

(5) Let P_{10} be of all polynomials of one variable of degree less than 10 with complex coefficients as a vector space over $\mathbb{C}$. Find all torsion factors and Betti numbers of the derivative.

(6) Express the commutative group $\mathbb{Z}^3/(f_1, f_2, f_3)$ as a direct sum of cyclic group where $f_1 = (4, 6, 9)$, $f_2 = (2, 4, 12)$, $f_3 = (4, 8, 16)$.

(7) Let $S = \mathbb{R}[x]$ where $\mathbb{R}$ is the real field. Express $S^3/(f_1, f_2, f_3)S^3$ in terms of the fundamental theorem where $f_1 = (x - 1, x, 0)$, $f_2 = (x, x - 1, 0)$, $f_3 = (0, 0, x - 1)$.

2.7   Rational Form and the Characteristic Polynomial

Let us define the notion of *characteristic polynomial* of a square matrix A as,

Definition The *characteristic polynomial* of a square matrix A is defined to be the determinant of $xI - A$. ∎

This definition is common in elementary linear algebra. We will postpone a deep discussion of it. For the time being, we will discuss only its mathematical properties.

Let us consider the following example,

Example 2.24: Let us study a simple case $M \cong R/(c)$ where $R = K[x]$ and $c = f(x) = x^m + a_1 x^{m-1} + \cdots + a_m$ with the linear operator A represented by the multiplication by x. We shall use $\{1, x, \ldots, x^{m-1}\} = \{v_1, v_2, \ldots, v_m\}$ as a basis for the *vector space* $K[x]/(f(x))$. Then we have $x = x \cdot 1 = A v_1$, $x \cdot x = A \cdot v_2 = x^2, \ldots, x \cdot x^{m-1} = A \cdot v_m = -\sum a_i x^{m-i} = \sum a_i v_{m-i+1}$. Therefore the matrix of A is similar to the following matrix

$$\tilde{A} = \begin{bmatrix} 0 & 0 & \cdot & \cdot & 0 & -a_m \\ 1 & 0 & \cdot & \cdot & \cdot & -a_{m-1} \\ 0 & 1 & \cdot & \cdot & \cdot & -a_{m-2} \\ \cdot & \cdot & \cdot & \cdot & & \cdot \\ 0 & 0 & \cdot & \cdot & 1 & -a_1 \end{bmatrix}.$$

The above is called the *invariant form* of A.

Furthermore, suppose we have matrix $\tilde{A}$ as above, then we may compute the *characteristic polynomial* $\chi_{\tilde{A}(x)}$ of $\tilde{A} = \det(xI - \tilde{A})$ as follows,

$$\det \begin{bmatrix} x & 0 & \cdot & \cdot & 0 & a_m \\ -1 & x & \cdot & \cdot & \cdot & a_{m-1} \\ 0 & -1 & \cdot & \cdot & \cdot & a_{m-2} \\ \cdot & \cdot & \cdot & \cdot & x & \cdot \\ 0 & 0 & \cdot & \cdot & -1 & x + a_1 \end{bmatrix}.$$

Let us multiply the mth row by x and add the result to the $(m - 1)$th row, and then multiply the new $(m - 1)$th row by x and add the result to the $(m - 2)$th row. We shall continue this process backward, the final result will be

$$\det \begin{bmatrix} 0 & 0 & \cdot & \cdot & 0 & c(x) \\ -1 & 0 & \cdot & \cdot & \cdot & \cdot \\ 0 & -1 & \cdot & \cdot & \cdot & \cdot \\ \cdot & \cdot & \cdot & \cdot & 0 & \cdot \\ 0 & 0 & \cdot & \cdot & -1 & x + a_1 \end{bmatrix}.$$

It is not hard to see that its characteristic polynomial is the original $\pm c(x)$.

Let us apply the process of finding the *Smith Normal Form* of the preceding matrix $xI - \tilde{A}$.

(1) We use the column operations to make the last column $[c(x), 0, \ldots, 0]^T$.
(2) Change the minus sign of the first $n - 1$ columns to positive.
(3) Interchange the first row with the second row, and then interchange the second row with the third row, and so on. Finally we have the following matrix

$$\begin{bmatrix} 1 & 0 & \cdot & \cdot & 0 & 0 \\ 0 & 1 & \cdot & \cdot & \cdot & 0 \\ 0 & 0 & \cdot & \cdot & \cdot & \cdot \\ \cdot & \cdot & \cdot & \cdot & 1 & 0 \\ 0 & 0 & \cdot & \cdot & 0 & c(x) \end{bmatrix}.$$

We conclude that if there is only one *torsion factor*, then it can be obtained by using the process of finding the *Smith Normal Form*. In fact the restriction to one torsion factor is not necessary. The same proposition works for any number of torsion factors. ∎

In general, given an $n \times n$ matrix A, according to the fundamental theorem, it is similar to the action by x on the torsion decomposition $R/(c_1) \oplus R/(c_2) \oplus \cdots \oplus R/(c_\ell)$. Then on each block, the action of x can be written in the *invariant form*. We may put everything together and write a similar form of A as

$$B = \begin{bmatrix} B_1 & 0 & \cdot & \cdot & 0 \\ 0 & B_2 & \cdot & \cdot & \cdot \\ \cdot & \cdot & \cdot & \cdot & \cdot \\ 0 & 0 & 0 & B_{\ell-1} & 0 \\ 0 & 0 & \cdot & \cdot & B_\ell \end{bmatrix}$$

where each B_j is in the invariant form as in the previous example. It may be called the *rational form* of A.

Let B be an $n \times n$ matrix and B_i be an $n_i \times n_i$ matrix for $i = 1, \ldots, \ell$ and $n = \sum_i n_i$. Then the characteristic polynomial $\chi_B(x)$ of B is $f(x) = \prod_i c_i(x)$, where $c_i(x)$ is computed in Example 2.26.

Using the discussions of the previous example, we shall deduce the following proposition,

Proposition 2.7.1: *Let A be a square matrix over a field. We apply the process to find the Smith Normal Form of $xI - A$. Then it is as follows,*

$$\begin{bmatrix} 1 & 0 & \cdot & \cdot & 0 & 0 \\ 0 & 1 & \cdot & \cdot & & 0 \\ 0 & 0 & \cdot & \cdot & & \cdot \\ \cdot & \cdot & \cdot & c_1(x) & 0 & 0 \\ \cdot & \cdot & \cdot & \cdots & \cdots & 0 \\ 0 & 0 & \cdot & \cdot & 0 & c_\ell(x) \end{bmatrix}$$

where only 1, and $c_i(x)$ are on the diagonal, all other places are 0. Furthermore, the number of 1 on the diagonal is $\sum(n_i - 1)$, and all $c_i(x)$ appear once and only once.

Proof. By selecting a different basis for the vector space K^n, we may find a new matrix B such that $B = C^{-1}AC$. It is easy to see that $\chi_B(x) = \det(xI - C^{-1}AC) = \det(C^{-1})\det(xI - A)\det(C) = \det(C^{-1}C)\det(xI - A) = \det(xI - A) = \chi_A(x)$. Therefore, we may select the basis so that the matrix is in the rational form. Then the theorem follows easily.   ∎

One important application is that any polynomial $c(x)$ can be made to be the *characteristic polynomial* of a matrix A. Therefore, if we can find approximate roots of a *characteristic polynomial* efficiently, then we can find approximations of the roots of any polynomial. That is precisely the way to do in computational algebra. For a detail of this process, see Section 6.3.

There is an interesting theoretic statement: the **Cayley–Hamilton theorem**: we have $\chi_A(A) = 0$ (cf. Section 2.10). Before we prove it in Section 10, let us consider the following example:

Example 2.25: Let

$$A = \begin{bmatrix} 1 & 2 \\ 3 & 4 \end{bmatrix}.$$

Then $\chi_A(x) = \det(xI - A) = x^2 - 5x - 2$. We have,

$$A^2 - 5A - 2I = \begin{bmatrix} 7 & 10 \\ 15 & 22 \end{bmatrix} - 5\begin{bmatrix} 1 & 2 \\ 3 & 4 \end{bmatrix} - 2\begin{bmatrix} 1 & 0 \\ 0 & 1 \end{bmatrix} = \begin{bmatrix} 0 & 0 \\ 0 & 0 \end{bmatrix}.$$ ∎

Example 2.26 (Computations of the Torsion and the Rational Form): Let us discuss a small example; let the field be $\mathbb{Q}$ and A be the following matrix,

$$A = \begin{bmatrix} 2 & -2 & 14 \\ 0 & 3 & -7 \\ 0 & 0 & 2 \end{bmatrix}$$

and

$$xI - A = \begin{bmatrix} x - 2 & 2 & -14 \\ 0 & x - 3 & 7 \\ 0 & 0 & x - 2 \end{bmatrix}.$$

Its Smith Normal Form is

$$\begin{bmatrix} 1 & 0 & 0 \\ 0 & x-2 & 0 \\ 0 & 0 & x^2 - 5x + 6 \end{bmatrix}.$$

The original generating set $[v_1, v_2, v_3]^T$ over $\mathbb{Q}[x]$ is modified by using elements in $\mathbb{Q}[x]$ to $[u_1, u_2, u_3]$. We have

$$1 \cdot u_1 = 0, \quad u_1 = 0, \quad (x-2)u_2 = 0, \quad xu_2 = 2u_2$$

$$(x^2 - 5x + 6)u_3 = 0 \quad \text{with } u_4 \equiv xu_3, \ xu_4 = 5u_4 - 6u_3.$$

Note that $\{u_2, u_3\}$ is a module generating set, $\{u_2, u_3, u_4\}$ is the vector space basis and the rational form of A is

$$\begin{bmatrix} 2 & 0 & 0 \\ 0 & 0 & -6 \\ 0 & 1 & 5 \end{bmatrix}.$$

We deduce easily that the torsion form of V is $\mathbb{Q}[x]/(x - 2) \oplus \mathbb{Q}[x]/(x^2 - 5x + 6)$.

Let us go back to the general problem. Once we get the Smith Normal Form over $K[x]$, we disregard the first few items on the diagonal which are units, the rest are the torsion factors, and from there we can easily deduce the rational form of A. One natural check of our computation is that the determinant of $xI - A$ is a polynomial of x with degree n. The Gaussian process can be realized by a sequence of multiplications of square matrices with determinants in $K \backslash \{0\}$. Therefore the degree in x will not change and stays as n. ∎

Exercises

(1) Find the number of non-isomorphic commutative groups of order 50.

(2) Let A be the following matrix with coefficients in $\mathbb{R}$,

$$A = \begin{bmatrix} 1 & 2 & 3 \\ 4 & 5 & 6 \\ 7 & 8 & 9 \end{bmatrix}.$$

Let $\mathbb{R}^3$, the real space, be considered as a $\mathbb{R}[x]$ module defined by

$$f(x) \cdot v = f(A) \cdot v.$$

What are the torsion factors and the Betti numbers?

(3) Write down the invariant matrix form of the action by x on the module $K[x]/(x^3 + x^2 + 1)$.

(4) Find the rational form of the following matrix A with coefficients in $\mathbb{R}$,

$$A = \begin{bmatrix} 0 & -1 & -1 \\ 0 & 0 & 0 \\ -1 & 0 & 0 \end{bmatrix}.$$

(5) Find the rational form of the following matrix A,

$$A = \begin{bmatrix} 1 & 2 & -4 & 4 \\ 2 & -1 & 4 & -8 \\ 1 & 0 & 1 & -2 \\ 0 & 1 & -2 & 1 \end{bmatrix}.$$

(6) Show that every Euclidean domain is a P.I.D.

(7) Show that every P.I.D. is a U.F.D.

(8) Construct vector space V, linear operators A and B such that A^2 is similar to B^2 while A is not similar to B.

(9) Prove that if $A^2(I - A) = A(I - A)^2 = 0$, then A is idempotent.

(10) If A, B, C, D are linear operators on a vector space V, and if both $A + B$ and $A - B$ are invertible, then there are linear operators X, Y such that

$$A = \begin{bmatrix} AX + BY = C \\ BX + AY = D \end{bmatrix}.$$

2.8 Torsion Decomposition and Elementary Decomposition

In the discussion of the theorem of the Smith Normal Form, we conclude in the theorem that the ideal generated by the left corner entry

a'_{11} will be uniquely determined by the original matrix A. Similarly, a theorem for the uniqueness of the remaining terms $c_2, \ldots, c_\ell$ is wanted. For this purpose, we need the following discussions about the *Chinese Remainder Theorem*. The following lemma is a common knowledge for P.I.D.,

Lemma 2.8.1: *Let the ring R be a P.I.D. Let m_i for $i = 1, \ldots, s$ a set of pairwise co-prime elements in R, i.e., $(m_i, m_j) = (1)$ for all $i \neq j$. Let $n_i = \prod_{j \neq i} m_j$. Then m_i, n_i are co-prime, i.e., $(m_i, n_i) = (1)$.*

Proof. Let us make an induction on the number s. If $s = 1, 2$, the lemma is trivial. Let us assume that it is true for $s - 1$. Without losing generality, we may assume that $i = 1$. Let $d = \prod_{j=2}^{s-1} m_j$. Then $(m_1, d) = (1)$, $n_1 = dm_s$. We have a, b such that

$$am_1 + bd = 1.$$

If $(m_1, n_1) \neq (1)$, let p be a non-unit factor of m_1, n_1. Then we have

$$am_s m_1 + bn_1 = m_s$$

hence $p | (m_1, m_s) = (1)$. A contradiction. Therefore we conclude $(m_1, n_1) = (1)$. ∎

The Chinese Remainder Theorem was known in the book *Sun Tzu's Arithmetic* in the 4th century A.D. The following is a variation of the Chinese Remainder Theorem,

Theorem 2.8.2 (Chinese Remainder Theorem I): *Let R be a P.I.D. and $c \in R$ is neither 0 nor a unit. Let the following be the prime decomposition of c (here we assume that it is known to the reader that P.I.D. $\Rightarrow$ U.F.D.),*

$$c = \delta \prod_{i=1} p_i^{n_i} \quad \delta \text{ unit}, p_i, p_j \text{ are non-associates if } i \neq j.$$

Then we have

$$R/(c) \approx \oplus_i R/(p_i^{n_i}).$$

Proof. Let $d = \delta \prod_{i=2} p_i^{n_i}$. Then clearly we have $c = p_1^{s_1} \cdot d$. We **claim** that

$$R/(c) \approx R/(p_1^{n_1}) \oplus R/(d).$$

Then we repeat the above process to decompose d. After finitely many repetitions, our theorem will be proved.

Because there is no common prime divisor of $p_1^{s_1}$ and d, we must have

$$(p_1^{s_1}, d) = (1).$$

Therefore there exist $\alpha, \beta \in R$ such that

$$\alpha p_1^{s_1} + \beta d = 1$$

(this is like the partition of unity in *topology*). Therefore we conclude that $p_1^{s_1}$ is a unit mod d, d is a unit mod $p_1^{s_1}$ and $r = r \cdot \alpha p_1^{s_1} + r \cdot \beta d$. We **claim** that the map $\chi : R \mapsto R/(p_1^{s_1}) \oplus R/(d)$ with $\chi(r) = (r \bmod (p_1^{s_1}), r \bmod (d))$ is onto and with $\text{Ker}(\chi) = (c)$. Let us show that χ is onto. Let $(\bar{r_1}, \bar{r_2}) \in R/(p_1^{s_1}) \oplus R/(d)$. Let $r = r_1 \beta d + r_2 \alpha p_1^{s_1}$. Then we have

$$r \bmod p_1^{s_1} = \overline{r_1 \beta d} \equiv \overline{r_1(1 - \alpha p_1^{s_1})} = \overline{r_1}.$$

Similarly we have

$$r \bmod d = \overline{r_2}.$$

Therefore χ is an onto map. Now let us show that the $\text{Ker}(\chi) = (c)$. Let $r \in \text{Ker}(\chi)$. Then we have

$$r \bmod(p_1^{s_1}) = 0$$

$$r \bmod(d) = 0.$$

Therefore we have $p_1^{s_1}, d | r$. Since these two factors are co-prime, we have $c = p_1^{s_1} \cdot d | r$ and $r \in \text{Ker}(\phi) = (c)$. $\blacksquare$

Remark: The original problem was to find *the smallest (non-negative) integer* n, such that $n \equiv 2(3)$, $n \equiv 3(5)$, $n \equiv 2(7)$. The answer is $n = 23$. In the preceding theorem, we only have that

$$\mathbb{Z}/(3 \cdot 5 \cdot 7) \approx \mathbb{Z}/(3) \oplus \mathbb{Z}/(5) \oplus \mathbb{Z}/(7).$$

The preceding theorem tells us that there is an element $\in \mathbb{Z}/(3 \cdot 5 \cdot 7)$ which will go to $(2, 3, 2) \in \mathbb{Z}/(3) \oplus \mathbb{Z}/(5) \oplus \mathbb{Z}/(7)$. In other words, we show the *existence* and *uniqueness* of solutions to a system of congruence equations. ■

Theorem 2.8.3 (Chinese Remainder Theorem II): *Let R be a P.I.D. and m_i for $i = 1, \ldots, s$ a set of pairwise co-prime elements in R, i.e., $(m_i, m_j) = (1)$ for all $i \neq j$. Then let $n_i = \prod_{j \neq i} m_j$, we have $(m_i, n_i) = (1)$, i.e., they are co-prime. Furthermore, let α_i satisfy*

$$\beta_i m_i + \alpha_i n_i = 1,$$

then $\alpha_i n_i$ satisfies the following system of equations $x_i \equiv 1 \pmod{m_i}$ and $x_i \equiv 0 \pmod{n_i}$. Furthermore let y_i, $i = 1, \ldots, s$, be any solution for the preceding system of equations, then the following system of equivalent equations $z = r_i \pmod{m_i}$ for $i = 1, \ldots, s$ can be solved by taking $z =$ the principal remainder of $\sum r_i y_i \pmod{\prod_i m_i}$.

Proof. We will have α_i such that

$$\beta_i m_1 + \alpha_i n_i = 1.$$

Then clearly $m_j | \alpha_i n_i$ for all $j \neq i$, and $m_i | n_j, \forall j \neq i$, henceforth, $\alpha_i n_i \equiv 1 \pmod{m_i}$. We conclude that $\alpha_i n_i$ satisfies the system of equations for x_i. Let $n' = \sum r_i y_i$. Then clearly we have $n' \equiv r_i \pmod{m_i}$. Our conclusion follows. ■

Example 2.27: The original problem was *the smallest (non-negative) integer n, such that $n \equiv 2(3)$, $n \equiv 3(5)$, $n \equiv 2(7)$.* The answer is $n = 23$. In the preceding theorem, we have to compute and find that

$$1(3 \cdot 5) + 1(3 \cdot 7) - 1(5 \cdot 7) = 1.$$

The preceding theorem tells us that there is an element $2(3 \cdot 5) + 3(3 \cdot 7) - 2(5 \cdot 7) = 23$ is a solution. Furthermore, 23 is a principle remainder. It is the solution to the residue equation of the problem. ■

Example 2.28 (Lagrange[12] Interpolation Formula): Given the values $v_1, v_2, \ldots, v_{n+1}$ of a polynomial $f(x) \in \mathbb{R}[x]$ with degree $\leq n$ at $n + 1$ distinct points $a_1, a_2, \ldots, a_{n+1}$.

Clearly that $\mathbb{R}[x]$ is a P.I.D. We have the following congruence relation; $f(x) = v_i \ (\text{mod}(x - a_i))$. According to the Chinese Remainder Theorem, it is enough to solve the system of equations $x_i \equiv 1 \ (\text{mod} \ \prod_{j \neq i}(x - a_j))$ and $x_i \equiv 0 \ (\text{mod} \ \prod_{j \neq k}(x - a_j)), \forall k \neq i$. Clearly we may take y_i to be $y_i = \frac{\prod_{j \neq i}(x - a_j)}{\prod_{j \neq i}(a_i - a_j)}$. Then the solution will be

$$f(x) = \sum_i v_i y_i = \sum_i \frac{v_i \prod_{j \neq i}(x - a_j)}{\prod_{j \neq i}(a_i - a_j)}. \qquad \blacksquare$$

Example 2.29 (RSA Encryption): It is the most populous *public key* encryption system today. RSA stands for Ron Rivest,[13] Adi Shamir[14] and Leonard Adleman,[15] who first publicly described it in 1978. It is an open way to transmit secret messages.

RSA involves a public key and a private key. The public key can be known to everyone and is used for encrypting messages. Messages encrypted with the public key can only be decrypted using the private key. The keys for the RSA algorithm are generated in the following way:

Choose two distinct large (may be with 200 digits) prime numbers p and q. For security purposes, the integers p and q should be chosen at random, and should be of similar bit-length. Prime integers can be efficiently found using a prime number test. Compute $n = pq$. The number n is used as the modulus for both the public and private keys, while the factorization $n = pq$ will be kept as a secret. Compute $\phi(n) = (p - 1)(q - 1)$, this number $\phi(n)$ will be kept secret, where ϕ is Euler's function.[16] Choose an integer e such that $1 < e < \phi(n)$ and the greatest common divisor of $(e, \phi(n)) = 1$, i.e., there is a number

---

[12]Lagrange, J. Italian mathematician. 1736–1813.
[13]Rivest, R. American cryptographer. 1947–.
[14]Shamir, A. Israeli cryptographer. 1952–.
[15]Adleman, L. American computer scientist. 1945–.
[16]Euler, L. Swiss mathematician, physicist, astronomer, logician and engineer. 1707–1783.

d such that

$$de + a\phi(n) = 1.$$

Then e is released as the public key exponent. The public key consists of the modulus n and the public (or encryption) exponent e.

Encryption Process: Alice transmits her public key (n, e) to Bob and keeps the private key (i.e., the number d) secret. Bob then wishes to send a message M to Alice.

He first turns M into an integer m. We restrict the length of m such that $0 < m < n$ by cutting the M into blocks. Then he creates a code $c = m^e \bmod(n)$.

Decryption Process: Alice can recover m from c by using her private key exponent d via computing

$$m = c^d \bmod(n).$$

We shall use the Chinese Remainder Theorem to show that Alice indeed recovers the message. It suffices to show that $m \equiv c^d \bmod(p)$ and $m \equiv c^d \bmod(q)$. Let us prove one of the equations, say the first one. It follows from *Fermat's little theorem* that $c^{p-1} = 1 \bmod(p)$ if $c \neq 0 \bmod(p)$. On the other hand, $c = 0 \bmod(p)$ iff $m = 0 \bmod(p)$, and $c = m \bmod(p)$ if $c = 0 \bmod(p)$. Let us assume that $c \neq 0 \bmod(p)$. We have $c^d \equiv (m^e)^d \equiv m^{de} \equiv m^{1-a(p-1)(q-1)} \equiv m \cdot (m^{(p-1)})^{-a(q-1)} \equiv m \cdot 1^{-a(q-1)} \equiv m \bmod(p)$. ∎

It follows from the fundamental theorem of finitely generated module M that M is isomorphic to $R/(c_1) \oplus R/(c_2) \oplus \cdots \oplus R/(c_\ell) \oplus R \oplus \cdots \oplus R$. It follows from the Chinese Remainder Theorem that M is isomorphic to $R/(p_1^{s_1}) \oplus R/(p_2^{s_2}) \oplus \cdots \oplus R/(p_t^{s_t}) \oplus R \oplus \cdots \oplus R$, where p_j are all primes. We have the following definition,

Definition (Elementary Factors): Let a finitely generated module M be isomorphic to $R/(p_1^{s_1}) \oplus R/(p_2^{s_2}) \oplus \cdots \oplus R/(p_t^{s_t}) \oplus R \oplus \cdots \oplus R$, where p_j are all primes (hence non-unit), then $(p_1^{s_1}), (p_2^{s_2}), \ldots, (p_t^{s_t})$ are called the *elementary factors* of M. ∎

The following definition will be used in the proof of the next theorem,

Definition (p-Submodule): Let R be a P.I.D. with p a prime element. The p-submodule $M(p)$ is defined to be

$$M(p) = \{m \in M;\ p^n m = 0 \text{ for some positive integer } n\}. \quad \blacksquare$$

Lemma 2.8.4: *Let R be a P.I.D., and M a finitely generated R-module with*

$$M = R/(p_1^{s_1}) \oplus R/(p_2^{s_2}) \oplus \cdots \oplus R/(p_t^{s_t}) \oplus R \oplus \cdots \oplus R.$$

(1) *Let N be the torsion module of M. Then we have*

$$N = R/(p_1^{s_1}) \oplus R/(p_2^{s_2}) \oplus \cdots \oplus R/(p_t^{s_t})$$

the part without the free copies of R.

(2) $M(p) = \oplus R/p_i^{s_i}$ *where p_i runs through all p_i which are associated with p.*

Proof. (1) It is easy to see the right-hand side is included in the left-hand side. Let us prove the other way. Let $\gamma = \alpha \oplus \beta \in N$ with $\alpha \in$ the right-hand side and $\beta \notin$ the right-hand side. Then γ, α are both of finite order, hence $\gamma - \alpha = \beta$ is of finite order. A contradiction.

(2) Similar to (1). $\quad \blacksquare$

Theorem 2.8.5 (Uniqueness of the Betti Number and the Elementary Factors): *Let the following be a decomposition of a finitely generated module M over a P.I.D. R,*

$$M \approx R/(p_1^{s_1}) \oplus R/(p_2^{s_2}) \oplus \cdots \oplus R/(p_t^{s_t}) \oplus R \oplus \cdots \oplus R.$$

Then the Betti number and the elementary factors $\{p_i^{s_i}\}$ are common for all decompositions.

Proof. Clearly we have uniquely

$$M = (\oplus_i M(p_i)) \oplus F = N \oplus F$$

where F is a free R-module. It is clearly $M/N = F$. Therefore F is unique for M. Its rank, the Betti number, is unique.

It suffices to show the further decomposition of $M(p_i)$ is unique. In other words, we may assume that $M = M(p_i)$ and prove the theorem.

Let us assume that $M = M(p)$. Let the following be one of the possible decompositions,

$$M = \oplus_{i=1}^{t} R/(p^{s_i}).$$

Let us assume that $s_i \geq 2$ for $i = 1, \ldots, q$ and $s_i = 1$ for all $q < i \leq t$, then we have

$$(p)M = \oplus_{i=1}^{q}(p)R/(p^{s_i}) = \oplus_{i=1}^{q} R/(p^{s_i-1}).$$

By induction on the $\sum s_i$, replacing M by $(p)M$, we conclude that the set $\{s_1, \ldots, s_q\}$ is unique. Let us consider the module M. It is easy to see that

$$M = (\oplus_{i=1}^{q} R/(p^{s_i})) \oplus (\oplus_{i=q+1}^{t} R/(p)R),$$

the second part is a free module over $R/(p)$ (it is easy to see that it is in fact a field and the direct sum of the field is a vector space over the field). Since the first part is common to all decomposition. Therefore the number $t - q$ must be common to all decompositions. ∎

From the above theorem, we know the uniqueness of the ideals $\{(p_i^{s_i})\}$. Since (c_j) are combinations of $(p_i^{s_i})$ with the further restriction that $c_1 | c_2 | \cdots | c_\ell$. We will use combinatorics to prove the following theorem,

Theorem 2.8.6 (Uniqueness of Torsion Factors): *Let M be a finitely generated module over a P.I.D. Then the torsion factors of M are unique up to multiplications by units.*

Proof. Let us consider the collection of all elementary factors $\{p_i^{s_i}\}$ where we may assume that any two p_i, p_j are either non-associate or identical (from all associated primes we only use one). We define the length $\ell(p_i)$ as the number of $p_i^{s_i}$ appears. In the following matrix, we arrange p_i as row with $\ell(p_1) \geq \ell(p_2)\ldots$, for every row, we arrange the power s_{ij} in a non-increasing order,

$$\begin{bmatrix} p_1 : & p_1^{s_{11}} & p_1^{s_{12}} & \cdots & p_1^{s_{1i}} \\ p_2 : & p_2^{s_{21}} & p_2^{s_{22}} & \cdots & p_2^{s_{2i}} \\ p_3 : & p_3^{s_{31}} & p_3^{s_{32}} & \cdots & p_3^{s_{3i}} \\ & \cdots & \cdots & & \\ p_t : & p_t^{s_{t1}} & p_t^{s_{t2}} & \cdots & p_t^{s_{ti}} \end{bmatrix}.$$

If $p_j^{s_{ji}}$ does not exist, then we define it to be 1. The only possible selection of $c_1|c_2|\cdots|c_\ell$ is that $c_\ell =$ the product of the first column, $c_{\ell-1} =$ the product of the second column, etc. Therefore our theorem is proved. ∎

Corollary (Uniqueness of the Smith Normal Form): *In the statement of the Smith Normal Form, all a'_{ii}'s are unique for all $1 \leq i \leq \ell$.*

Proof. Let N be the submodule generated by all columns. Then it trivially follows from the preceding theorem. ∎

Exercises

(1) Find the smallest positive integer x which satisfies $x \equiv 2$ $\mod(3)$, $x \equiv 3 \mod(5)$, $x \equiv 5 \mod(8)$.

(2) Find the number of all non-isomorphic commutative groups of order p^3 where p is a prime number.

(3) Find all similarity classes of 3×3 matrices with elementary factors a power of $x - a$.

(4) Compute the elementary factors to show that the following two matrices are similar over the rational field $\mathbb{Q}$,

$$\begin{bmatrix} 0 & -4 & 85 \\ 1 & 4 & -30 \\ 0 & 0 & -3 \end{bmatrix}, \quad \begin{bmatrix} 2 & 2 & 1 \\ 0 & 2 & -1 \\ 0 & 0 & 3 \end{bmatrix}.$$

(5) Find the number of non-isomorphic commutative group of order 100.

(6) For a small example of the RSA encryption system, let $p = 41$, $q = 23$, then $n = p \cdot q = 943$. Let $e = 7$ and announce the public keys $(943, 7)$. List the two smallest positive integers which are the possible decryption key.

2.9 Jördan Canonical Form. Differential Equations

In this section, we will be given a finite dimensional vector space over an *algebraically closed field* which is defined as

Definition A field K is said to be *algebraically closed* if one of the following equivalent conditions are satisfied,

(1) Every non-constant polynomial $f(x) \in K[x]$ has a *root* α in K, i.e., $f(\alpha) = 0$.
(2) The only irreducible polynomials are linear ones.
(3) Every non-constant polynomial $f(x) \in K[x]$ can be written as a product of linear ones, i.e.,

$$f(x) = \prod_i l_i(x), \quad \text{where } l_i(x) \text{ are linear polynomials.} \quad \blacksquare$$

Remark: The definition of algebraically closed field is common. The reader is referred to any book on algebra about it. $\quad \blacksquare$

Remark: One important fact in algebra, the so-called *Fundamental Theorem of Algebra* of Gauss is that the field C of complex numbers is an algebraically closed field. We will refer the reader to any book on algebra. $\quad \blacksquare$

Definition Let A be an $n \times n$ matrix that acts on an n-dimensional vector space V. Let U be a subspace of V. If A induces a map $A : U \mapsto U$, then we say U is an *invariant subspace* of A. $\quad \blacksquare$

Let V be a finite dimensional vector space over an algebraically closed field K, say C. Let A be the matrix of a linear operator. It follows from the theorem on the elementary factors that A can be represented by x acting on $\oplus_i K[x]/(p_i(x)^{s_i})$. Note that in the case of an algebraically closed field, each $p_j(x)$ is linear and $(p_j^{s_j}) = ((x - \alpha_j)^{s_j})$. We will have the following definition and theorem,

Definition (Jördan[17] Block): A square matrix J is said to be a Jördan block, if it is of the following form with some λ's on the diagonal and 1's below the diagonal,

$$J = \begin{bmatrix} \lambda & 0 & 0 & 0 & 0 \\ 1 & \lambda & 0 & 0 & 0 \\ 0 & 1 & \lambda & 0 & 0 \\ \cdot & \cdot & \cdot & \cdot & \cdot \\ 0 & 0 & 0 & 1 & \lambda \end{bmatrix}.$$

∎

Theorem 2.9.1 (Jördan Canonical Form): *Let A be the matrix form of a linear operator with respect to some basis. If (the minimal polynomial, see Theorem 2.10.6) $c_\ell(x)$ splits completely into a linear product, then A is uniquely similar to the following matrix with the Jördan blocks on the diagonal, and 0's elsewhere and the uniqueness up to a reordering of the Jördan blocks,*

$$J = \begin{bmatrix} J_1 & 0 & 0 & 0 & 0 \\ 0 & J_2 & 0 & 0 & 0 \\ 0 & 0 & \cdot & 0 & 0 \\ \cdot & \cdot & \cdot & \cdot & \cdot \\ 0 & 0 & 0 & 0 & J_h \end{bmatrix}.$$

Especially, if the ground field is algebraically closed, then the requirement of the theorem is always satisfied.

Proof. According to the existence of the elementary factors, the matrix A is represented by x acting on the module $\oplus_i K[x]/(p_i(x)^{s_i})$. If the field K is algebraically closed, all irreducible polynomials $p_i(x)$ are of the form $(x - \lambda_i)$. If $c_\ell(x)$ splits completely into a linear product, then all elementary factors are powers of linear polynomials. Let us consider the action of x on one component $K[x]/((x - \lambda_i)^{s_i})$. Let us take a basis of it as $b_0 = 1$, $b_1 = (\bar{x} - \lambda_i), \ldots, b_j = (\bar{x} - \lambda_i)^j, \ldots, b_{s_i-1} = (\bar{x} - \lambda_i)^{s_i-1}$. Then we

---

[17] Jördan, C. French mathematician. 1838–1922.

clearly have $xb_0 = \bar{x} = b_1 + \lambda_i b_0, \ldots, xb_j = \bar{x}(\bar{x} - \lambda_i)^j = (\bar{x} - \lambda_i)^{j+1} + \lambda_i b_j = b_{j+1} + \lambda_i b_j, \ldots, xb_{s_i-1} = (\bar{x} - \lambda_i)^{s_i} + \lambda_i b_{s_i-1} = \lambda_i b_{s_i-1}$ where $(\bar{x} - \lambda_i)^{s_i} = 0$ in $K[x]/((x - \lambda_i)^{s_i})$. The action of x has the matrix form of a Jördan block as follows,

$$J = \begin{bmatrix} \lambda_i & 0 & 0 & 0 & 0 \\ 1 & \lambda_i & 0 & 0 & 0 \\ 0 & 1 & \lambda_i & 0 & 0 \\ \cdot & \cdot & \cdot & \cdot & \cdot \\ 0 & 0 & 0 & 1 & \lambda_i \end{bmatrix}.$$
∎

Remark: A matrix can be diagonalized iff all Jördan blocks are 1×1 iff the minimal polynomial $c_\ell(x)$ splits completely into linear product and has no multiple root. ∎

Corollary 2.9.2: *A matrix A is similar to an upper-triangular matrix iff the minimal polynomial splits into a product of linear factors.*

Proof. ($\Longrightarrow$) Clearly we have $\chi_A(\lambda) = \prod \lambda - a_{ii} = \prod_j c_j(\lambda)$, and $c_\ell(\lambda)$ is split into a product of linear forms.

($\Longleftarrow$): The Jördan canonical form of A is in lower-triangular form. If we reverse the order of the basis $\{e_1, e_2, \ldots, e_n\}$ to $\{e_n, e_{n-1}, \ldots, e_1\}$, then a lower-triangular matrix becomes an upper-triangular matrix. ∎

If the field K is algebraically closed, then all polynomials split into products of linear factors, therefore any square matrix is similar to an upper-triangular form. Note that in the complex case, it happens. Note that any given non-constant complex polynomial $f(x)$ can be realized as a characteristic polynomial of a suitable matrix A over complex numbers. Furthermore, A can be upper-triangularized, and thus the roots of $f(x)$ are numbers on the diagonal. This provides a way to find the roots of any given polynomial $f(x)$. The only problem is to find an effective way to upper-triangularize a square matrix A over a complex number. In the real number case, there is a similar problem. We will provide the discussions about these methods later in Section 6.3 of this book.

Remark: In the proof of the Jördan canonical form, we only require that all elementary factors of A are powers of a linear polynomial. Therefore even if the field is not algebraically closed, as long as all elementary factors of A are powers of a linear one, then we still have the Jördan canonical form of a matrix A. ∎

Example 2.30 (System of Differential Equations): Let us consider a system of differential equations $x' = Ax$ where x is a vector of complex functions $x = [f_1(t), \ldots, f_n(t)]^T$ with n entries, A an $n \times n$ complex matrix. It follows from the preceding theorem that A can be expressed as

$$A = M^{-1}JM$$

where M is invertible and J is a matrix in the Jördan canonical form. In other words, we have $x' = M^{-1}JMx$ or $(Mx)' = Mx' = JMx$, since M is a constant matrix. Let $y = Mx$, then we have $y' = Jy$. A standard set of equations are of the following form,

$$\frac{dy_1}{dt} = \lambda y_1 + y_2$$
$$\cdots$$
$$\frac{dy_{s-1}}{dt} = \lambda y_{s-1} + y_s$$
$$\frac{dy_s}{dt} = \lambda y_s.$$

Let t be the variable, then $y_s = ce^{\lambda t}$, $y_{s-1} = f_{s-1}(t)e^{\lambda t}, \ldots, y_1 = f_1(t)e^{\lambda t}$, where $f_i(t)$ is a polynomial in t of degree $s - i$ for $i = s - 1, \ldots, 1$ with $f_i'(t) = f_{i+1}(t)$ with $f_s = c$. Once we find y, then we find x by

$$x = M^{-1}y.$$ ∎

Remark: Recall the definition of *similarity*. Two A, B matrices are similar, i.e., there is invertible $n \times n$ matrix C such that $A = C^{-1}BC$. It is the same to say two matrices are similar iff they are the matrix representations of the same linear operator ϕ with respect to two bases.

The design of a microchip used in a computer involves solving a system of differential equations $x' = Ax$. The above Example 2.30 is very useful in industry. ∎

Given a square matrix A, it is difficult to find the elementary factors even if the field K is algebraically closed. It is equivalently difficult to compute the eigenvalues. Let us assume that we solve the just-mentioned difficult problems, then the following method is interesting for a constructive and directive finding of the Jördan form.

Example 2.31 (Filippov[18]): Given a square matrix $A_{n \times n}$ over C, knowing its eigenvalues, we want to find its Jördan form directly. Note that the property of a basis $\{w_1, \ldots, w_n\}$ with respect to which the matrix A is in the Jördan canonical form is that the basis $\{w_1, \ldots, w_n\}$ will be separated into several blocks such that within each block, we have

$$Aw_i = \lambda_i w_i \quad \text{or} \quad Aw_i = \lambda_i w_i + w_{i+1}.$$

We shall try to achieve this.

(1) Let $B = A - \lambda I$ where λ is a variable. The equation $\det(B) = 0$ can be expressed as a polynomial equation in λ of degree n. Let us select a root λ_0 for the equation (usually it is difficult!). Then we have $\det(A - \lambda_0 I) = 0$. Therefore the matrix $B = A - \lambda_0 I$ is singular. Suppose we can find the Jördan canonical form of B, then $A = B + \lambda_0 I$, and A's Jördan canonical form can be found easily. So we shall try to find the Jördan canonical form of a singular matrix A.

(2) Note that since A is singular, we have $\text{rank}(A) = r < \dim(A) = n$. Let $U = $ column space$(A) = $ image(A) which is a smaller dimensional space, and A restricts to a linear map of U to U, by the induction on the dimension (we note that the Jördan canonical form trivially exists for 1-dimensional vector space, and we may start a mathematical induction on dimension!), we know that the restriction of A to U, $A|_U$, has a Jördan canonical

---

[18]Filippov, A. F. Russian mathematician. 1923–2006.

form, which means there is a basis $\{b_1, \ldots, b_r\}$ which satisfies
the condition in the first paragraph. We wish to extend it to the
whole vector space.

(3) Let the nullspace of A be W, $U \cap W = P$, and $W = P \oplus S$. Let
$\dim(P) = p$, and $\dim(S) = n - r - p$. Note that p is precisely
the number of blocks with the eigenvalues 0. The end vector of
those blocks is with eigenvalue 0. Certainly a vector v is in the
nullspace iff $Av = 0 = 0v$, i.e., it is a vector with eigenvalue 0.
It is not hard to see that P is generated by all end vectors of
all blocks with eigenvalue 0. We wish to add a new vector at
the beginning of these blocks. Let $\{\omega_1, \ldots, \omega_p\}$ be the starting
vectors at these p blocks. Since each of them $\in U$, therefore it
must be a combination of the columns of the matrix. So we have
$\omega_i = Ay_i$ for some suitable y_i for $i = 1, \ldots, p$.

(4) The nullspace of A always have dimension $n - r$. We may write
$W = P \oplus S$, where S is of dimension $n - r - p$. Let us find a
basis $z_1, \ldots, z_{n-r-p}$ for it.

To sum up, we have r vectors w_i, p vector y_i and $n - r - p$ vectors
z_i. We **claim** that A is in the Jördan form with respect to the above-
mentioned vectors.

Proof of the Claim: The only thing we have to prove is that the
vectors $\{w_i\}, \{y_i\}, \{z_i\}$ are linearly independent, then they form a
basis. Assume the converse. Let the following be a linear equation
among them,

$$\sum a_i w_i + \sum b_i y_i + \sum c_i z_i = 0.$$

Apply the matrix A to the above equation, we have $Az_i = 0$, Ay_i
is the first vector w_j which appears in a block with eigenvalue 0,
so it will not appear in the first summation sign after applying A.
Therefore we conclude $b_i = 0$ for all i, or we have

$$\sum a_i w_i + \sum c_i z_i = 0.$$

The above is an equation among the linearly independent vectors
$\{w_i\}, \{z_i\}$, therefore all a_i, c_i must be zero. We finish the proof of the
claim.   ∎

Example 2.32: Let us consider the following two examples for Filippov's method.

(a) Let

$$A = \begin{bmatrix} 1 & 1 & 1 \\ 0 & 0 & 0 \\ 0 & 0 & 0 \end{bmatrix}.$$

Since A is singular, and the column space U is generated by c_1, we have $\dim(U) = 1$, and the restriction of A to U is 1×1 matrix (1), it is in the Jördan form. Furthermore, $W \cap U = \{0\}$, we have the Jördan form of A as follows,

$$\begin{bmatrix} 1 & 0 & 0 \\ 0 & 0 & 0 \\ 0 & 0 & 0 \end{bmatrix}.$$

(b) Let

$$A = \begin{bmatrix} 9 & 0 & 0 & 8 & 8 \\ 0 & 1 & 0 & 8 & 8 \\ 0 & 0 & 1 & 0 & 0 \\ 0 & 0 & 0 & 1 & 0 \\ 0 & 0 & 0 & 0 & 9 \end{bmatrix}.$$

Since A is non-singular, we shall define $A' = A - I$ which is singular as follows,

$$A' = \begin{bmatrix} 8 & 0 & 0 & 8 & 8 \\ 0 & 0 & 0 & 8 & 8 \\ 0 & 0 & 0 & 0 & 0 \\ 0 & 0 & 0 & 0 & 0 \\ 0 & 0 & 0 & 0 & 8 \end{bmatrix}.$$

Now we start reducing the sizes of the matrix to 1×1. Clearly the column space is generated by the column vectors c_1, c_4, c_5. Let the restriction of the mapping A to the column space be B. Let us use c_1, c_4, c_5 as a basis for the column space. Then we have $Bc_1 = 8c_1$,

$Bc_4 = 8c_1$, $Bc_5 = 8c_1 + 8c_5$. The matrix B can be written as

$$B = \begin{bmatrix} 8 & 8 & 8 \\ 0 & 0 & 0 \\ 0 & 0 & 8 \end{bmatrix}.$$

Let the three-column vectors of B be named as b_1, b_2, b_3. Then the column space of B is generated by b_1, b_3. Let the matrix of the restriction of B to this column space be C. Then $Cb_1 = 8b_1$, $Cb_3 = 8b_1 + 8b_3$. Now let us select $b_3, 8b_1$ as a basis, then the matrix form of C is

$$C = \begin{bmatrix} 8 & 0 \\ 1 & 8 \end{bmatrix}.$$

We shall consider $C - 8I = D$. Let the two columns of D be d_1, d_2. Then we have $D(d_1) = D(d_2) = 0$. In other words, the column space of D is generated by d_1. We reduce the mapping to the subspace generated by d_1, and we find the eigenvalue is 0. Now we start going backward. We extend the block with 0 eigenvalue by one more vector which is $[1, 0]^T$. We use the vectors $[1, 0]^T$, $[0, 1]^T$ as a basis for $C = D + 8I$ and find the above expression for C as its Jördan form. We shall go one step backward. Now let us select $b_1 - b_2$, b_3, $8b_1$ as basis, then the form of B is

$$B = \begin{bmatrix} 0 & 0 & 0 \\ 0 & 8 & 0 \\ 0 & 1 & 8 \end{bmatrix}.$$

Once we have this Jördan form, we will go backward to find the Jördan form for A. Note that the first column $b_1 - b_2 = c_1 - c_4 = [0, -8, 0, 0, 0]^T = A[1, 0, 0, -1, 0]^T$, we shall name $y_1 = [1, 0, 0, -1, 0]^T$ and $z_1 = [0, 0, 1, 0, 0]^T$. With respect to $z_1, y_1, c_1 - c_4, c_5, 8c_1$, the form of A' is

$$\begin{bmatrix} 0 & 0 & 0 & 0 & 0 \\ 0 & 0 & 0 & 0 & 0 \\ 0 & 1 & 0 & 0 & 0 \\ 0 & 0 & 0 & 8 & 0 \\ 0 & 0 & 0 & 1 & 8 \end{bmatrix}$$

and the Jördan form of $A = A' + I$ is the following

$$\begin{bmatrix} 1 & 0 & 0 & 0 & 0 \\ 0 & 1 & 0 & 0 & 0 \\ 0 & 1 & 1 & 0 & 0 \\ 0 & 0 & 0 & 9 & 0 \\ 0 & 0 & 0 & 1 & 9 \end{bmatrix}.$$

∎

Exercises

(1) Find, by inspection, the Jördan canonical form of the following matrix

$$A = \begin{bmatrix} 1 & 1 & 1 \\ 0 & 2 & 0 \\ 0 & 0 & 3 \end{bmatrix}.$$

(2) Find, by inspection, the Jördan canonical form of the following matrix

$$A = \begin{bmatrix} 1 & 1 & 1 \\ 1 & 1 & 1 \\ 1 & 1 & 1 \end{bmatrix}.$$

(3) Find the Jördan canonical form of the following matrix

$$A = \begin{bmatrix} 0 & 1 & 1 \\ 0 & 0 & 0 \\ 0 & 0 & 0 \end{bmatrix}.$$

(4) Find the Jördan canonical form of the following matrix

$$A = \begin{bmatrix} 0 & 1 & 1 \\ 1 & 1 & 0 \\ 0 & 0 & 0 \end{bmatrix}.$$

(5) Solve the following system of differential equations,

$$\begin{bmatrix} x' \\ y' \\ z' \end{bmatrix} = \begin{bmatrix} 0 & 1 & 1 \\ 0 & 0 & 1 \\ 0 & 0 & 0 \end{bmatrix} \begin{bmatrix} x \\ y \\ z \end{bmatrix}.$$

(6) Show that the matrix $\left[\begin{smallmatrix} 0 & 1 \\ -1 & 0 \end{smallmatrix}\right]$ is not similar to $\left[\begin{smallmatrix} \lambda_1 & 0 \\ 0 & \lambda_2 \end{smallmatrix}\right]$ nor to $\left[\begin{smallmatrix} \lambda_1 & 0 \\ 1 & \lambda_1 \end{smallmatrix}\right]$ over $\mathbb{R}$.

(7) **(Jördan Canonical Form for Real $\mathbb{R}$ Matrices):** We know from algebra that the only irreducible polynomials in $\mathbb{R}[x]$ are either (1) linear i.e., of the form $c(x) = x - \lambda$ or (2) quadratic i.e., of the form $c(x) = (x - a)^2 + b^2$ with $b \neq 0$. Let us consider the linear case. Show that it is identical with the Jördan block in the complex case. Show that if it is the quadratic case, let $I = 2 \times 2$ identity matrix, and D be defined as follows,

$$D = \begin{bmatrix} a & b \\ -b & a \end{bmatrix}.$$

Then the Jördan block is as follows,

$$J = \begin{bmatrix} D & 0 & \cdot & 0 & 0 \\ I & D & \cdot & 0 & 0 \\ 0 & I & \cdot & 0 & 0 \\ \cdot & \cdot & \cdot & \cdot & \cdot \\ 0 & 0 & \cdot & I & D \end{bmatrix}$$

where in the above form 0 is the 2×2 zero matrix.

2.10 Eigenvalue, Eigenvector and the Cayley–Hamilton Theorem

The concepts of *eigenvalues and eigenvectors* originated in the study of *Euler's principle axis of rotations in three-dimensional space* $\mathbb{R}^3$ (see *Euler's rotation theorem* below) by *Lagrange.*[19] This is the *Fourier Series* (see Section 4.2, *Trigonometric Series*) of solutions of the following equations,

$$\frac{\partial^2 g}{\partial x^2} = -n^2 g(x).$$

---

[19]Lagrange, J. Italian mathematician. 1736–1813.

Fourier[20] gave the solutions as

$$g(x) = \pm \sin(nx) \text{ or } \pm \cos(nx).$$

The solutions of the above equation, which can be thought of as $Lg = -n^2 g$ with L a linear operator, had many usages in *physics* and *engineering*. Later on, it was discovered to be an important concept in linear algebra. At the turn of 20th century, German mathematicians Hilbert,[21] von Helmholtz,[22] Courant,[23] etc. used the German term *eigen* in the study of eigenvalue and eigenvector, and it became standard in several languages.

In this section, we assume that the reader is familiar with the elementary properties of determinants by expansions in the first chapter. Let us formulate the following definition for a finite dimensional vector space.

Definition (Eigenvalues and Eigenvectors): Let A be a linear operator on K^n, if for some $0 \neq v \in K^n$, we have

$$Av = \lambda v$$

where $\lambda \in K$, then we say λ is an eigenvalue of A, and v is an eigenvector associated with λ. ∎

Example 2.33: Let us consider the rotations in the 3-dimensional space $\mathbb{R}^3$ which fixes the origin $[0, 0, 0]^T$. All rotations as specified above can be expressed as a matrix A. Then the vectors from the origin to the north pole or south pole are two eigenvectors with the corresponding eigenvalue 1 (see below). ∎

Example 2.34: Let A be a square matrix and λ_0 be an eigenvalue of A. Let $V_0 = \{v : Av = \lambda_0 v\}$ (v may be a 0 vector). Then V_0 is an invariant subspace under A. ∎

---

[20] Fourier, J. French mathematician and physicist. 1768–1830.
[21] Hilbert, D. German mathematician. 1862–1943.
[22] von Helmholtz, H. German mathematician, physicist and psychologist. 1821–1894.
[23] Courant, R. German–American mathematician. 1888–1972.

How to find the eigenvalues and eigenvectors? Let us consider the following example:

Example 2.35: Let

$$A = \begin{bmatrix} 1 & 2 \\ 2 & 1 \end{bmatrix}.$$

A systematic way of finding the eigenvalues and eigenvectors of A is as follows; starting with the eigenequation $Av = \lambda v \Rightarrow Av = \lambda I v \Rightarrow (A - \lambda I)v = 0$, then the matrix $(A - \lambda I)$ has a non-zero element $v \in$ the kernel $\Leftrightarrow$ its determinant is 0. We have the following *characteristic equation,*

$$\det(A - \lambda I) = \lambda^2 - 2\lambda - 3 = 0.$$

Solving the above equation, we have $\lambda = -1, 3$. Replacing λ by $-1, 3$ in the matrix $(A - \lambda I)$, and then computing its nullspace, we find $[1, -1]^T, [1, 1]^T$ as associated eigenvectors. ∎

Definition (Characteristic Equation): We define the characteristic polynomial, $\chi_A(\lambda)$, of a square matrix A as $\det(\lambda I - A)$ with λ a variable. We define the characteristic equation of a square matrix A as $\chi_A(\lambda) = 0$. ∎

Definition (Trace): Given an $n \times n$ matrix $A = (a_{ij})$. We define the trace of A, $\text{trace}(A) = \sum a_{ii}$. Let the characteristic polynomial of A be $\chi_A(\lambda) = \lambda^n + a_1 \lambda^{n-1} + \cdots$. It is easy to see that $a_1 = (-1)\text{trace}(A)$. ∎

Proposition 2.10.1: *If two A, B matrices are similar, then* $\chi_A(\lambda) = \chi_B(\lambda)$.

Proof. There is an invertible matrix C such that $A = C^{-1}BC$. Therefore $\chi_A(\lambda) = \det(A - \lambda I) = \det(C^{-1}BC - \lambda I) = \det(C^{-1}BC - C^{-1}\lambda I C) = \det(C^{-1}(B - \lambda I)C) = \det C^{-1} \chi_B(\lambda) \det(C) = \chi_B(\lambda)$. ∎

Corollary 2.10.2: *If two A, B matrices are similar, then* $\text{trace}(A) = \text{trace}(B)$. ∎

Example 2.36: The above proposition is not sufficient. Let $A = \begin{bmatrix} 1 & 0 \\ 1 & 1 \end{bmatrix}$, $B = \begin{bmatrix} 1 & 0 \\ 0 & 1 \end{bmatrix}$. Then $\chi_A(\lambda) = (\lambda - 1)^2 = \chi_B(\lambda)$, and A, B are not similar to each other. ∎

Definition (Minimal Polynomial): The minimal polynomial of the non-zero matrix A is a non-zero monic polynomial $f(x)$ of minimal degree such that $f(A)$ is the identical zero matrix. We use $m_A(x)$ to denote the minimal polynomial of A. ∎

Proposition 2.10.3: *If A represents a linear operator on V, $V = U \oplus W$ where U, W are invariant subspaces under A, then $\chi(A) = \chi(A|_U)\chi(A|_W)$, and $m(A) = lcm(m(A|_U), m(A|_W))$.*

Proof. Let U be represented by the first m coordinates and W be represented by the last $n - m$ coordinate. Then A is represented by

$$A = \begin{bmatrix} B & 0 \\ 0 & C \end{bmatrix}$$

where B is an $m \times m$ matrix, C is an $(n - m) \times (n - m)$ matrix, the two 0's indicate two $m \times (n - m)$ and $(n - m) \times m$ zero matrices. Clearly we have

$$A - \lambda I = \begin{bmatrix} B - \lambda I & 0 \\ 0 & C - \lambda I \end{bmatrix}.$$

The first part of our proposition follows. The second part is easy. ∎

Proposition 2.10.4: *Assume that the ground field K is algebraically closed. Assume that the matrix A is represented by x on $K[x]/c(x)$. If $c(x)$ is a power of a linear polynomial, then the characteristic polynomial and the minimal polynomial of the matrix A are $\pm c(\lambda)$ and $c(x)$, respectively.*

Proof. Let $c(x) = (x - \lambda_i)^\ell$. Then it follows from the Jördan canonical form that

$$A \cong \begin{bmatrix} \lambda_i & 0 & 0 & 0 & 0 \\ 1 & \lambda_i & 0 & 0 & 0 \\ 0 & 1 & \lambda_i & 0 & 0 \\ \cdot & \cdot & \cdot & \cdot & \cdot \\ 0 & 0 & 0 & 1 & \lambda_i \end{bmatrix}.$$

Then clearly we have $\chi_A(\lambda) = \det(A - \lambda I) = \pm(\lambda - \lambda_i)^\ell = \pm c(\lambda)$. As for the minimal polynomial $m(x)$, it is clear that all non-zero polynomials of degree less than the degree $c(x) = \ell$ can be reinterpreted as a linear combination between a basis $1, x, \ldots, x^{\ell-1}$. Therefore it cannot be 0. We conclude that $m(x)$ is trivially $c(x)$. ∎

Without the assumption of algebraically closeness of the ground field K, the above proposition is still true. In fact, we have the following proposition,

Proposition 2.10.5: *The characteristic polynomial and the minimal polynomial of the matrix A represented by x on $K[x]/(c(x))$, where $c(x)$ is a non-zero monic polynomial, are $\pm c(\lambda)$ and $c(x)$, respectively.*

Proof. Let $c(x)$ be monic and $c(x) = x^\ell + \alpha_1 x^{\ell-1} + \cdots + \alpha_\ell$. Since the characteristic polynomial is common among all similar matrices, we may select a basis to express A in its invariant form as

$$A \cong \begin{bmatrix} 0 & 0 & 0 & 0 & -\alpha_n \\ 1 & 0 & 0 & 0 & -\alpha_{n-1} \\ 0 & 1 & 0 & 0 & -\alpha_{n-2} \\ \cdot & \cdot & \cdot & \cdot & \cdot \\ 0 & 0 & 0 & 1 & -\alpha_1 \end{bmatrix}.$$

Then the characteristic polynomial $\chi_A(\lambda)$ of A is the following,

$$\chi_A(\lambda) = \det \begin{bmatrix} \lambda & 0 & 0 & 0 & \alpha_n \\ -1 & \lambda & 0 & 0 & \alpha_{n-1} \\ 0 & -1 & \lambda & 0 & \alpha_{n-2} \\ \cdot & \cdot & \cdot & \cdot & \cdot \\ 0 & 0 & 0 & -1 & \lambda + \alpha_1 \end{bmatrix}.$$

We shall multiply the nth row by λ to add to the $(n-1)$th row to kill the term λ, and repeat the procedure of multiplying the (resulting) $(n-1)$th row by λ to add to the $(n-2)$th row and continue this way backward. The final determinant is of the following form,

$$\chi_A(\lambda) = \det \begin{bmatrix} 0 & 0 & 0 & 0 & c(\lambda) \\ -1 & 0 & 0 & 0 & \cdot \\ 0 & -1 & 0 & 0 & \cdot \\ \cdot & \cdot & \cdot & \cdot & \cdot \\ 0 & 0 & 0 & -1 & \lambda + \alpha_1 \end{bmatrix}.$$

Now we expand the determinant according to the first column, then the second column, ..., it it easy to see that

$$\chi_A(\lambda) = \pm c_\lambda.$$

As for the minimal polynomial, the proof is the same as the preceding proposition. ∎

Theorem 2.10.6: *If A is represented by x acting on $\oplus_{i=1}^{\ell} K[x]/$ $(c_i(x))$ where $c_i(x)$ are monic polynomials with $c_1(x)|c_2(x)|\cdots|c_\ell(x)$. Then we have*

$$\chi_A(\lambda) = \prod_{i=1}^{\ell} c_i(x)$$

$$m_A(x) = c_\ell(x).$$

Proof. Clearly the characteristic polynomial $\chi_A(\lambda)$ is independent of a field extension. We may extend the field F to an algebraically closed field Ω. In other words, we may assume K is algebraically closed. Then the elementary factors are all linear. Our theorem about characteristic polynomials are correct for each elementary components. The statement about characteristic polynomial follows by noticing that the product of elementary factors equal the product of torsion factors.

For the minimal polynomial $m(x)$, since $c_i(x)|c_\ell(x)$ $\forall i$, then $c_\ell(x)$ inducing zero maps on all components of the torsion decomposition, hence $c_\ell(x)$ is zero on the whole vector space. On the other hand, any proper factor of $c_\ell(x)$ will not be zero for the last component in the torsion decomposition. Therefore it cannot be zero identically. Thus we conclude that $c_\ell(x)$ is the minimal polynomial. ∎

Theorem 2.10.7 (Cayley[24]–Hamilton[25]): *We always have* $\chi_A(A) = 0$.

Proof. Since $m_A(A) = 0$ and $m_A(\lambda)|\chi_A(\lambda)$, then we have $\chi_A(A) = 0$.   ■

Remark: Cayley only showed his theorem for 2×2 and 3×3 cases. The general theorem was proved by Frobenius.[26]   ■

Lemma 2.10.8: *Let J be an $m \times m$ Jordan block as follows,*

$$J = \begin{bmatrix} \lambda_i & 0 & 0 & 0 & 0 \\ 1 & \lambda_i & 0 & 0 & 0 \\ 0 & 1 & \lambda_i & 0 & 0 \\ \cdot & \cdot & \cdot & \cdot & \cdot \\ 0 & 0 & 0 & 1 & \lambda_i \end{bmatrix}.$$

Then J^k is of the following form

$$J^k = \begin{bmatrix} \lambda_i^k & 0 & 0 & 0 & 0 \\ c_1^k \lambda^{k-1} & \lambda_i^k & 0 & 0 & 0 \\ c_2^k \lambda^{k-2} & c_1^k \lambda^{k-1} & \lambda_i^k & 0 & 0 \\ \cdot & \cdot & \cdot & \cdot & \cdot \\ c_{m-1}^k \lambda^{k-m+1} & \cdot & \cdot & c_1^k \lambda^{k-1} & \lambda_i^k \end{bmatrix}.$$

Proof. Exercises.   ■

Example 2.37 (Euler Rotation Theorem): Every non-trivial rotation in $(2n + 1)$-dimensional real space has a pair of north pole and south pole, where it is a rotation around the poles. We will assume the critical properties of a *rotation* A center at the origin: (1) $\det A = 1$, (2) $AA^T = I$, (3) the rotations keep the length of any vector, which implies that the eigenvalues, if they exist, of A are ± 1, i.e., $Av = \lambda v$ implies $\lambda = \pm 1$, that is to say the length of Av equals

[24]Cayley, A. British mathematician. 1821–1895.
[25]Hamilton, W. Irish mathematician. 1805–1865.
[26]Frobenius, F. German mathematician. 1849–1917.

the length of v. Note that $Av = v$ gives us the north pole and south pole if such an equation defines exactly two points.

We give two proofs.

(1) We **claim** that 1 is an eigenvalue of A. The characteristic equation of A, $\chi_A(\lambda)$, is a polynomial of degree $2n + 1$. From *Calculus*, it follows from the mean value theorem that a polynomial of odd degree must have a real root, therefore, A must have an eigenvalue, and it must be ± 1. If it is $+1$, then we establish our claim. Assume the other way. We assume that -1 is an eigenvalue. In the following,

$$\chi_A(\lambda) = -\lambda^{2n+1} + a_1\lambda^{2n} + \cdots + a_{2n}\lambda + a_{2n+1}$$

if we set $\lambda = 0$, the left-hand equals $\det(A)$ which is 1, the right-hand equals a_{2n+1}. Henceforth we have

$$\chi_A(\lambda) = -\lambda^{2n+1} + a_1\lambda^{2n} + \cdots + a_{2n}\lambda + 1$$
$$= -(\lambda^{2n} + \cdots + b_{2n-1}\lambda + b_{2n})(\lambda + 1).$$

It is easy to see that $-b_{2n} \cdot 1 = 1$. Therefore $b_{2n} = -1$. From the elementary theory of real functions, we know that the true complex roots of a real polynomial comes in conjugating pairs, $a + bi, a - bi$, and their equation is $\lambda^2 - 2a + a^2 + b^2$ with constant term $a^2 + b^2 > 0$. Since we know that $b_{2n} = -1 < 0$. We conclude that not all roots of $\chi_A(\lambda)/(\lambda + 1)$ cannot be true complex roots. As arguing before, we know that the product of all constants is -1. Therefore there must be some linear factors with negative constant term. Henceforth, there will be some positive real root which must be 1.

(2) We **claim** that $\det(A - I) = 0$. If it is true, then we know the nullspace of $A - I$ is not $\{0\}$. Let $0 \neq v \in Null(A - I)$. Then we have $(A - I)v = 0$ or $Av = Iv = v$, and 1 is an eigenvalue.

We have $A^T(A - I) = I - A^T$, $\det(I - A^T)^T = \det(I - A)$. Then we have $\det(A^T(A - I)) = \det(A^T)\det(A - I) = \det(A)\det(A - I) = \det(A - I) = \det(I - A)$. Since the dimension is odd, then the last equation is $\det(I - A) = (-1)^{2n+1}(\det(A - I)) = -\det(A - I)$. Over the real field, we have $\det(A - I) = 0$ as claimed. ∎

Proposition 2.10.9: *A matrix A can be diagonalized $\Leftrightarrow$ the minimal polynomial splits into a product of distinct linear monic factors.*

Proof. ($\Rightarrow$) Trivial.

($\Leftarrow$) Since $m_A(\lambda) = c_\ell(\lambda)$ which is a multiple of $c_i(\lambda)$. Therefore $c_i(\lambda)$ splits into a product of distinct linear monic factors for all i. It follows that all elementary factors are linear and A can be diagonalized. ∎

Exercises

(1) Let A be a 2×2 matrix with real entries. Show that $\text{trace}(A) = 0 = \text{trace}(A^2) \Leftrightarrow A^2 = 0$.

(2) Let V be a subspace of dimension $n - 1$ of an n-dimensional vector space U. Let ϕ be an endomorphism of U (i.e., $\phi(U) \subset U$) which fixes every element in V. We know that there is an element α such that $\phi(u) = au + t_u$, where a is common to every u and $t_u \in V$. If $\alpha \neq 1$, show that α is an eigenvalue of ϕ.

(3) Let A, B be two $n \times n$ matrices over K. Let λ be an eigenvalue for AB, then show that λ is an eigenvalue for BA.

(4) Let A, B be two similar operators on a finite dimensional vector space V. Show that $\text{trace}(A) = \text{trace}(B)$.

(5) Diagonalize over $\mathbb{C}$

$$\begin{bmatrix} 1 & -i & i \\ i & 2 & -2 \\ -i & -2 & 2 \end{bmatrix}.$$

(6) Show that

$$\begin{bmatrix} 1 & 5 & 0 \\ 0 & 2 & 7 \\ 0 & 0 & 3 \end{bmatrix}$$

can be diagonalized over the rational field $\mathbb{Q}$.

(7) Count the number of non-similar complex 3×3 matrices with 1 as the only eigenvalue.

(8) Use the Cayley–Hamilton theorem to find the monic minimal polynomial for the Fibonacci matrix

$$\begin{bmatrix} 1 & 1 \\ 1 & 0 \end{bmatrix}.$$

(9) Prove Lemma 2.10.8.

2.11 Simultaneously Diagonalizable Matrices

Later on we will learn that every hermitian operator acting on a finite dimensional complex vector space is diagonalizable. Matrices A, B can be simultaneously diagonalizable, then they are commutative, i.e., $AB = BA$. On the other hand, we will prove in this section that if matrices $A_1, \ldots, A_m$ are diagonalizable each individually, and they are all commutative, then they can be diagonalizable simultaneously.

Example 2.38: One of the fundamental rules of matrices is that their multiplications are non-commutative. For instance, we have

$$\begin{bmatrix} 0 & 1 \\ 0 & 1 \end{bmatrix}\begin{bmatrix} 1 & 0 \\ 1 & 0 \end{bmatrix} = \begin{bmatrix} 1 & 0 \\ 1 & 0 \end{bmatrix} \neq \begin{bmatrix} 0 & 1 \\ 0 & 1 \end{bmatrix} = \begin{bmatrix} 1 & 0 \\ 1 & 0 \end{bmatrix}\begin{bmatrix} 0 & 1 \\ 0 & 1 \end{bmatrix}.$$

All matrices above cannot be diagonalized simultaneously. Some naturally occurred linear operators (which may act on infinite dimensional vector spaces) are not commutative. For instance, let $\frac{\partial}{\partial x}$ and x be two operators acting on differentiable functions on the real line, we have $(\frac{\partial}{\partial x}x - x\frac{\partial}{\partial x})(f) = f = If$. Therefore these two operators are not commutative, hence they cannot be diagonalized simultaneously. This is the kernel of Heisenberg's uncertainty principle (see Section 4.5). ∎

We have the following theorem for the finite dimensional vector spaces,

Theorem 2.11.1: *Let $\{A_i\}$ be a finite set of commutative diagonalizable square matrices. Then there is a basis for the vector space V, such that all A_i are in diagonal form with respect to this basis.*

Proof. We make an induction with respect to the dimension of the vector space V. If $n = \dim(V) = 1$. Then the theorem is clearly true. Furthermore, if all A_i are constant matrices, i.e., $A_i = c_i I$, then the theorem is clearly true. Let us assume that A_1 is not a constant matrix. Let the characteristic values of A_1 be $c_1, \ldots, c_\ell$. Then $V = \oplus V_{c_i}$, where $V_{c_i} = \{v : A_1 v = c_i v\}$, the eigenspace of A_1 corresponding to the eigenvalue c_i. It is easy to see that for any $v \in V_{c_i}$, we always have

$$A_1 A_j(v) = A_j A_1(v) = c_i A_j(v).$$

Therefore $A_j(v) \in V_{c_i}$, i.e., V_{c_i} is invariant subspace for all A_j. Clearly the minimal polynomial of the restriction of A_j to V_{c_i} is a factor of the minimal polynomial of A_j which is with only linear factors since A_j is diagonalizable. Therefore the restriction is with only linear factors. Hence it can be diagonalizable. We have all restrictions of all A_j for all j can be diagonalizable. It works for all components V_{c_i}. So they all can be diagonalizable. ∎

Exercises

(1) Let A and B be linear operators such that $AB - BA$ commutes with A. Show that $A^k B - B A^k = k A^{k-1}(AB - BA)$.

(2) Find all matrix B that diagonalize the matrix A,

$$A = \begin{bmatrix} 1 & 4 \\ 0 & 2 \end{bmatrix}.$$

(3) Use the previous problem to find A^m for all m.

(4) If $1, 0$ are eigenvalues of A. Show that $1, 0$ are eigenvalues of A^2.

(5) Show that

$$\begin{bmatrix} 1 & 3 \\ 0 & 2 \end{bmatrix}, \quad \begin{bmatrix} 1 & 0 \\ 3 & 2 \end{bmatrix}$$

can be separably diagonalized while they cannot be diagonalized simultaneously.

(6) Show that

$$\begin{bmatrix} 1 & 0 \\ 0 & 2 \end{bmatrix}, \quad \begin{bmatrix} 1 & 8 \\ 3 & 2 \end{bmatrix}$$

can be separably diagonalized while they cannot be diagonalized simultaneously.

(7) Can the following matrices be simultaneously diagonalized?

$$\begin{bmatrix} 1 & 0 \\ -2 & 3 \end{bmatrix}, \quad \begin{bmatrix} 1 & 0 \\ -1 & 2 \end{bmatrix}.$$

Chapter 3

Determinants and Multilinear Algebras

3.1 Further Discussion of the Determinant. Volume in $\mathbb{R}^n$

This is a continuation of Section 3 in Chapter 1. The definition there using summations is concrete but not practical. The number of terms under the summation sign in the definition of determinant is the number of permutations of n objects, $n!$. Let $n = 30$. Then $30! \approx 2^{108}$ which is beyond the memory of a supercomputer. If we feed the computer sequentially, and if the computer is superfast, it will make 10^{12} multiplications per second; note that in every year there are 3×10^7 seconds. So every year there are 3×10^{19} computations. Let us use the formula $2^{10} \approx 10^3$, it requires $2^{108} \approx 10^{32.4}$ computations. So it needs 10^{13} years of computation time. Note that the age of the universe is 2×10^{10} years. To compute a 30×30 determinant by expanding the sequential, we will require more than the age of the universe using a fast computer to find the results. It is an impossible job. Let alone that we compute the determinant of $1,000,000 \times 1,000,000$ matrix routinely. What we are sure of is that if $n \geq 30$, then the elementary definition of the determinants is not computable. Its significance is in the theoretical understanding of determinants. We call for new definitions for other purposes.

Note that we have proved three important propositions for the *determinant* in Section 3 of Chapter 1. We shall prove that the three properties of the *determinant* mentioned there uniquely defined the function *determinant* in the following theorem.

In this chapter, we are given a ring (i.e., a commutative ring with identity $1 \neq 0$). We take the module $V = R^n$. We write the determinant of A as $\det A$, or $\det[v_a, \ldots, v_n]$ where $v_i = [a_{1i}, a_{2i}, \ldots, a_{ni}]^T$ is the ith column vector of A. We need the following lemma to proceed,

Lemma 3.1.1: *If one of the column vectors v_i's is 0, then any multilinear function f on n vector variables $[v_1, v_2, \ldots, v_n]$ must be 0, i.e., $f(A) = 0$.*

Proof. Let $A = [v_1, \ldots, 0, \ldots, v_n]$. Then $f(A) = f([v_1, \ldots, 0 + 0, \ldots, v_n]) = f(A) + f(A)$. Therefore $f(A) = 0$. ∎

Theorem 3.1.2 (Definition of Determinant by Axioms):

There is only one function f mapping from $\overbrace{V \times V \times \cdots \times V}^{n} \mapsto R$ which satisfies the following three axioms;

(1) *Multilinear Property: For any i, the function $f(c_1, \cdot, c_i, \cdot, c_n)$ is a linear function in the vector variable c_i with all other c_j fixed.*
(2) *Alternating Property: For any $i \neq j$, if $c_i = c_j$, then $f(c_1, \ldots, c_n) = 0$.*
(3) *Unimodule Property: $f(I) = 1$.*

Proof. By the propositions of Section 3, Chapter 1, we know that the determinant function satisfies the above conditions (1)–(3). Hence such functions exist. We claim that there is only one such function. Suppose that there are two functions $f([c_1, \ldots, c_n])$ and $g([c_1, \ldots, c_n])$ which satisfy all conditions (1)–(3). Let $A = [c_1, \ldots, c_n]$. We claim that $f([c_1, \ldots, c_n]) = g([c_1, \ldots, c_n])$ always, i.e., $h = (f - g) = 0$. Note that h satisfies the first two conditions, and $h(I) = 0$. Let A be any $n \times n$ matrix. We **claim** that $h(A) = 0$ always. This will show that $f = g$.

Let us prove the **claim**. We apply a sequence of column operations to A. If we apply the first elementary column operation to interchange the ith and jth column vectors, then by the first and second conditions, we have the following computations,

$$0 = h([c_1, \ldots, c_i + c_j, \ldots, c_i + c_j, \ldots, c_n])$$
$$= h([c_1, \ldots, c_i, \ldots, c_i, \ldots, c_n]) + h([c_1, \ldots, c_j, \ldots, c_j, \ldots, c_n])$$
$$+ h([c_1, \ldots, c_i, \ldots, c_j, \ldots, c_n]) + h([c_1, \ldots, c_j, \ldots, c_i, \ldots, c_n])$$
$$= h([c_1, \ldots, c_i, \ldots, c_j, \ldots, c_n]) + h([c_1, \ldots, c_j, \ldots, c_i, \ldots, c_n]).$$

Therefore we conclude

$$h(c_1, \ldots, c_i, \ldots, c_j, \ldots, c_n]) = -h(c_1, \ldots, c_j, \ldots, c_i, \ldots, c_n]),$$

the value of h will change sign. If we apply the second elementary column operations by multiplying the ith column by a non-zero number a, then $h(A) = a^{-1}h(A')$. If we multiply ith column by a number b and add the result to a distinct column, then the value will not change. Finally we get the reduced column echelon form of A, which is either I or the last column 0. In the first case, by our assumption of $h(I) = 0$. In the second case, by the preceding lemma, the value of h is zero again. The sequence of column operation is reversible. Hence the value $h(A) = 0$.

Since $h(A) = 0$ for any A, we must have $h = 0$, i.e., $f = g$. ∎

By the duality, we may change rows to columns everywhere in the axiomatic definition of determinant. By the uniqueness of the determinant, these two definitions must be the same.

By the preceding theorem, the three conditions above are *axioms* of the determinant. In fact a computer uses the axioms to compute a determinant over a field K. Given an $n \times n$ matrix A, the computer will scan the first row to find the first non-zero item (if it cannot find a non-zero term, then the determinant is zero), and switch it to the leftmost (it uses no time to switch), and multiply the first column by a_{11}^{-1} to make the first term 1, and then we use the first column to clear out all the first row. The number of multiplications is $n \times (n-1)$. Then it takes no time to clear out the first column. We work on the $(n-1) \times (n-1)$ after we delete the first column and the first row. The total number of multiplications to achieve the reduced column echelon form is

$$\sum_{i=1}^{n-1}(n-i+1)(n-i) \approx n^3/3.$$

In the case that the reduced column echelon form has a zero column, its determinant is zero. Otherwise, its determinant is the product of -1 (the first kind of elementary column operation) and a^{-1} (the second kind of elementary column operation).

For a 30×30 matrix, the number of computations for the determinant by the axioms is $30^3/3 = 9 \cdot 10^3$, which can be handled by a small computer in less than one second.

Laplace[1] Theorem

Given a matrix A, we have the following definitions of (i,j)th minor $M_{i,j}$ and (i,j)th cofactor $A_{i,j}$,

Definition Let $A = (a_{ij})$ be an $n \times n$ matrix. We define the (i,j)th *minor* M_{ij} as the $(n-1) \times (n-1)$ matrix that results from A by removing the ith row and the jth column. We define the (i,j)th *cofactor* A_{ij} as $(-1)^{i+j}\det(M_{ij})$.   ∎

We have the following theorem due to Laplace,

Theorem 3.1.3 (Definition of Determinant by Induction):
The Laplace formula is

$$\delta_{ik}\det(A) = \sum_{j}a_{ij}A_{kj} = \sum_{j}a_{ji}A_{jk}$$

where δ_{jk} is the Kronecker δ with value 1 if $j = k$, otherwise 0.

Proof. If $n = 1, 2$, the formula is obvious. We shall use induction and assume the formula is true for all smaller n. There are two cases depending on $i = k$. We first handle the case that $i = k$, i.e., $\delta_{ik} = 1$ and $\det(A) = \sum_{j}a_{ji}A_{ji} = \sum_{j}a_{ij}A_{ij}$. In fact these two equations are dual, and we shall prove only the first equation.

---

[1]Laplace, P. French mathematician and astronomer. Napoleon Bonaparte was his student. 1749–1828.

We will prove that the three axioms of determinant are satisfied by $f([c_1, \ldots, c_n]) = \sum_j a_{ji} A_{ji}$. Then it must be $\det(A)$. Select any ℓth column vector c_ℓ, if $\ell = i$, then f is linear in the components of the vector c_ℓ, therefore it is linear in c_ℓ. If $\ell \neq i$, then each $A_{i,j}$ is linear in the component of c_ℓ. Therefore f is linear in c_ℓ.

We want to prove that if $c_\ell = c_k$ with $\ell \neq k$ (we may assume that $\ell < k$), then $f = 0$. If $\ell \neq i$, $k \neq i$, then all $A_{ji} = 0$, hence $f = 0$. If $\ell = i$ and $k \neq i$, now we want to show that $\sum_j a_{ji} A_{ji} = 0$. By the induction hypothesis, $A_{ji} = (-1)^{i+j} \sum_k a_{rk} (A_{ij})_{rk}$. We compute the terms involving $a_{ji} a_{rk} = a_{ri} a_{jk}$, and we find they cancel out. Hence the total sum is zero.

It is easy to check that $f(I) = 1$. Therefore we have $f = $ determinant.

Furthermore it is easy to see that $\sum_j a_{ij} A_{kj} = f([c_1, \ldots, c_i, \ldots, c_i, \ldots, c_n]) = 0$ if $i \neq k$. Thus the second case $i \neq k$ is proved.

We have shown that $\sum_j a_{ji} A_{ji}$ satisfies the axioms of $\det(A)$, therefore $\det(A) = \sum_j a_{ji} A_{ji}$ with fixed i. ∎

Definition The *adjoint matrix*, $\operatorname{adj}(A)$, of A is the transpose of the matrix of cofactors, i.e.,

$$\operatorname{adj}(A) = (A_{ij})^T. \qquad ∎$$

Remark: Later on in Chapter 4 of the inner product, we will have a totally different definition of adjoint. It is very unfortunate! ∎

Proposition 3.1.4: *If A is a triangular matrix (i.e., $a_{ij} = 0$ for all $i > j$ or alternatively for all $i < j$), then*

$$\det(A) = \prod_{i=1}^{n} a_{ii}.$$

Proof. Assume that $a_{ij} = 0$ for all $i < j$, otherwise we take a transpose. We may expand by the first column, i.e.,

$$\det(A) = \sum_j a_{1j} A_{1j}.$$

By our assumption, $a_{12} = a_{13} = \cdots = a_{1n} = 0$, we conclude that $\det(A) = a_{11}A_{11}$. It follows from the mathematical induction on the size of matrices that $A_{11} = \prod_{i=2} a_{ii}$. Our proposition follows. ∎

Theorem 3.1.5: *We always have* $A \cdot \operatorname{adj}(A) = \det(A)I$.

Proof. It follows from the Laplace formula. Routine. ∎

Theorem 3.1.6: *A matrix A is invertible $\Leftrightarrow \det(A)$ is a unit. If $\det(A)$ is a unit, then $A^{-1} = \det(A)^{-1}(A_{ji})$.* ∎

Proof. We have $A \cdot \operatorname{adj}(A) = \det(A)I$. Therefore if $\det(A)$ is a unit, then $A^{-1} = (\det(A))^{-1} \operatorname{adj}(A)$.

On the other hand, if A is invertible, then there is a matrix B such that $AB = BA = I$. Therefore $\det(A)\det(B) = 1$, i.e., $\det(A)$ is a unit. ∎

Proposition 3.1.7 (Cramer's[2] Rule): *Let $A = (a_{ij})$ be an invertible $n \times n$ matrix, and the following system of linear equations,*

$$Ax = b$$

where $x = (x_1, \ldots, x_n)^T$, $b = (b_1, \ldots, b_n)^T$ are vectors. Then we have

$$x_i = \frac{1}{\det A} \cdot \det \begin{bmatrix} a_{11} & \cdot & b_1 & \cdot & \cdot & a_{1n} \\ \cdot & & \cdot & & \cdot & \cdot \\ a_{i1} & \cdot & b_i & \cdot & \cdot & a_{in} \\ \cdot & & \cdot & & \cdot & \cdot \\ a_{n1} & \cdot & b_n & \cdot & \cdot & a_{nn} \end{bmatrix}$$

where the vector $b = (b_1, \ldots, b_n)^T$ replaces the ith column of A.

Proof. From $Ax = b$ and A non-singular, we conclude that $x = A^{-1}b$. It follows from the preceding theorem that $x = \det(A)^{-1}(A_{ji})b$.

---

[2]Cramer, G. Genevan mathematician. 1704–1752.

We expand the matrix product $(A_{ji})b$, and find the ith entry x_i is precisely the ith term of the expansion of the above equation. ∎

The above definition is still non-computable because we have to inductively compute A_{ij} which involve $(n-1)!$ many terms. We have to do n further multiplications, and end up working on $n!$ terms again. However, it does help us to understand many problems. We will later have some usage of the preceding definition. For computational purpose, we need the following abstract definition of permutation matrices,

Definition A matrix $A = (a_{ij})$ is said to be a permutation matrix if on every column there is one entry 1 with the rest of the entries 0, and on every row there is one entry 1 with the rest of the entries 0. We always have $Ae_i = e_{i_j}$ for some i_j. Therefore it represents the permutation σ with $\sigma(i) = i_j$. Sometimes we write $A = A_\sigma$. ∎

Lemma 3.1.8: *Let A be a matrix that represents a permutation σ. Then we have $\det(A) = \text{sign}(\sigma)$.*

Proof. Elementary. ∎

Remark: For determinants, the definition could be given by expansion or by induction or by axioms. In other words, all three definitions of the determinant are equivalent. ∎

Example 3.1 (Signed n-Volume): Let $\mathbb{R}$ be the real number field. Let $v_1, v_2, \ldots, v_n$ be n vectors in $V(= \mathbb{R}^n)$, an n-dimensional real space.

Since $\det([v_1, v_2, \ldots, v_n])$ is the unique function of n vector variables which satisfies the three axioms of the definition of the determinant if we can show that the n-dimensional volume, $\text{vol}([v_1, v_2, \ldots, v_n])$, satisfies the three axioms of the definition of determinant, then it must be the determinant.

Instead of presenting the n-dimensional arguments in detail, we will illustrate the 2-dimensional case as follows. Let v_1, v_2 be two

vectors in the following picture

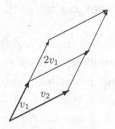

Clearly the area (i.e., 2-volume), $\text{vol}[2v_1, v_2]$, of the parallelogram formed by $2v_1, v_2$ will be double the area, $\text{vol}[v_1, v_2]$, i.e.,

$$\text{vol}[2v_1, v_2] = 2\,\text{vol}[v_1, v_2]$$

formed by v_1, v_2. If we have $v_1 = u_1 + u_2$, then the relation is kept by the projection to a line which is perpendicular to the line determined by v_2, i.e., $\text{proj}(v_1) = \text{proj}(u_1) + \text{proj}(u_2)$ which are precisely the three heights of the three parallelograms formed by $(v_1, v_2), (u_1, v_2), (u_2, v_2)$ with the common side of v_2 as the basis. Thus $\text{vol}[v_1, v_2] = \text{vol}[u_1, v_2] + \text{vol}[u_2, v_2]$. This shows essentially that axiom (1) is satisfied.

Given any vector, we have a definition of *upper* and *lower* half-planes (which works in general, given any $n-1$ linearly independent and ordered vectors $v_1, v_2, \ldots, v_{n-1}$, then the whole space $\mathbb{R}^n$ is separated into *upper* and *lower* half-spaces). If v_2 is in the upper half-plane, then we define the area to be positive, if v_2 is in the lower-plane, then the area is negative, and if v_2 lies on the line determined by v_1, then the area is 0.

With the above understanding of area, it is easy to see that if $v_1 = v_2$, then the area is 0. Thus we have checked axiom (2).

As for the last axiom (3), it is our general definition of area that a square with each side 1 is of area 1.

Therefore we conclude that the 2-dimensional area is the determinant of 2×2 matrices over the real number field $\mathbb{R}$. In general, we may argue along the same line to show that the n-dimensional volume of the parallelopiped span by n vectors $v_1, \ldots, v_n$ is the determinant of an $n \times n$ matrix with column vectors $v_1, \ldots, v_n$ over the real number field R. ∎

Example 3.2 (Gaussian Operations and det): The first and third definitions of the determinant are non-computable, then the definition by axioms comes to the rescue. Let us take a simple example,

$$\det(A) = \det \begin{bmatrix} 1 & 2 & 3 \\ 4 & 5 & 6 \\ 7 & 9 & 8 \end{bmatrix} = \det[c_1, c_2, c_3].$$

It follows from the three axioms that $\det[c_1, c_2, c_3] = \det[c_1, c_2 - 2c_1, c_3 - 3c_1]$. Therefore we have,

$$\det(A) = \det \begin{bmatrix} 1 & 0 & 0 \\ 4 & -3 & -6 \\ 7 & -5 & -13 \end{bmatrix} = \det[r_1, r_2, r_3]^T.$$

Notice that r_1, r_2, r_3 are row vectors. Furthermore we have $\det(A) = \det[r_1, r_2, r_3]^T = \det[r_1, r_2 - 4r_1, r_3 - 7r_1]^T$. We have

$$\det(A) = \det \begin{bmatrix} 1 & 0 & 0 \\ 0 & -3 & -6 \\ 0 & -5 & -13 \end{bmatrix} = \det[s_1, s_2, s_3]^T.$$

Then we have $\det[s_1, s_2, s_3]^T = -3 \det[s_1, -(1/3)s_2, s_3]^T$. Or we may write down as follows,

$$\det(A) = -3 \det \begin{bmatrix} 1 & 0 & 0 \\ 0 & 1 & 2 \\ 0 & -5 & -13 \end{bmatrix} = -3 \det[s_1, s_2', s_3]^T.$$

Then we have

$$\det(A) = -3 \det[s_1, s_2', s_3 + 5s_2']^T$$
$$= (-3)(-3) \det[s_1, s_2', (-1/5)(s_3 + 5s_2')]^T$$

as follows,

$$\det(A) = 9 \det \begin{bmatrix} 1 & 0 & 0 \\ 0 & 1 & 2 \\ 0 & 0 & 1 \end{bmatrix} = 9 \det[d_1, d_2, d_3]$$

where d_1, d_2, d_3 are the three-column vectors. Clearly we have $\det(A) = 9 \det[d_1, d - 2, d_3 - 2d_2] = 9 \det I = 9$. This way of computation is no different from the computation of the *echelon* form of the matrix A. The way of computation of the expansions of definition (1) or (2) involves $n!$ terms where each involves $n-1$ multiplications, so totally there are $(n-1)(n!)$ multiplications. While the last way of using Gaussian operations where we keep a record of interchanging two distinct vectors (we change the sign of the determinant if necessary), factor a non-zero factor r from a row (or column) (we multiply the determinant by r), the total number of multiplications to find the determinant this way is $\approx n^3/3$. There is a substantial saving if n is large. ∎

Example 3.3: Let us consider the following matrix A over Z. We shall use Gaussian operations to find its determinant,

$$A = \begin{bmatrix} 1 & 1 & 1 & 3 \\ 1 & 2 & 0 & 4 \\ 1 & 0 & 1 & 0 \\ 2 & 0 & 1 & 0 \end{bmatrix}.$$

Let us multiply the first row by -2 and add to the second row, and then expand according to the second column. We get

$$\det(A) = -1\det\begin{bmatrix} -1 & -2 & -2 \\ 1 & 1 & 0 \\ 2 & 1 & 0 \end{bmatrix}.$$

Now expand according to the third column, we have

$$\det(A) = 2\det\begin{bmatrix} 1 & 1 \\ 2 & 1 \end{bmatrix}.$$

It is easy to conclude that $\det(A) = -2$. ∎

Example 3.4: Let us consider the following matrix A over Z,

$$A = \begin{bmatrix} 1 & 1 & 1 & 3 \\ 1 & 0 & 0 & 4 \\ 1 & 0 & 1 & 0 \\ 2 & 0 & 1 & 0 \end{bmatrix}.$$

Then

$$\det(A) = -\det\begin{bmatrix} 1 & 0 & 4 \\ 1 & 1 & 0 \\ 2 & 1 & 0 \end{bmatrix} = -4\det\begin{bmatrix} 1 & 1 \\ 2 & 1 \end{bmatrix} = 4.$$

Thus $\det(A)$ is computed. ∎

Exercises

(1) Show that $\det(A)$ is the same for expansion by ith column and for expansion by jth row.

(2) Let $n > 1$. Show that the determinant of an $n \times n$ real matrix $A = (a_{ij})$ with $a_{ij} = ij$ has determinant 0.

(3) Compute the following determinant,

$$\det(A) = -\det \begin{bmatrix} 3 & 1 & 1 & 1 \\ 1 & 3 & 1 & 1 \\ 1 & 1 & 3 & 1 \\ 1 & 1 & 1 & 3 \end{bmatrix}.$$

(4) Find the area of the pentagon of the five vertices $(1, 2)$, $(4, 1)$, $(5, 3)$, $(3, 7)$, $(2, 6)$.

(5) A linear operator A on an n-dimensional vector space is said to be nilpotent if $A^m = 0$ for some $m \geq 0$. Prove that if A is nilpotent, then $A^n = 0$.

(6) Let a matrix A represent a permutation σ. Show that $\det(A) = \text{sign}(\sigma)$.

(7) Give a complete argument for Example 3.1.

(8) Prove Lemma 3.1.8.

3.2   Cauchy–Binet Formula

There are many more interesting geometric concepts other than just volume. For instance, the lengths and areas in $\mathbb{R}^3$. In Chapter 1, we have studied in $\mathbb{R}^n$ the m-dimensional volumes $\mathbb{A} = \mathbb{A}[v_1, \ldots, v_m]$ of the parallelopiped spanned by m linearly independent vectors $\{v_1, \ldots, v_m\}$ as $\sqrt{\det(A^T A)}$ where $A = [v_1, \ldots, v_m]$ is an $n \times m$ matrix.

Let us recall the following theorem from Section 1.5.

Theorem 3.2.1: *We always have* $\det(A^T A) = \mathbb{A}^2$. ∎

We may generalize the above $\det(A^T A)$ to the computation of $\det(A \cdot B)$ where A is an $m \times n$ matrix, and B is an $n \times m$ matrix, with $n \geq m$. Let N, M be two sets of cardinalities n, m respectively.

Then there are $= \binom{n}{m}$ many subsets of N with precise m elements. Let the set of subsets of N with m elements be S. We have the following theorem,

Theorem 3.2.2 (Cauchy[3]–Binet[4] Theorem): *Let us use the notation of the preceding paragraph. Let us assume the ring R is a field K. Let A be an $m \times n$ matrix and B an $n \times m$ matrix, then we have*

$$\det(A \cdot B) = \sum_s \det(A_s) \det(B_s)$$

where A_s is the $m \times m$ matrix that consists of the m columns determined by s. Similarly, B_s is the $m \times m$ matrix that consists of the m rows determined by s.

Proof. Let us write

$$A = \begin{bmatrix} a_{11} & a_{12} & a_{13} & \cdots & a_{1n} \\ a_{21} & a_{22} & a_{23} & \cdots & a_{2n} \\ & \cdots & \cdots & & \\ a_{m1} & a_{m2} & a_{m3} & \cdots & a_{mn} \end{bmatrix} = \begin{bmatrix} r_1 \\ r_2 \\ \cdot \\ r_m \end{bmatrix}$$

and

$$B = \begin{bmatrix} b_{11} & b_{12} & \cdots & b_{1m} \\ b_{21} & b_{22} & \cdots & b_{2m} \\ b_{31} & b_{32} & \cdots & b_{3m} \\ & \cdots & \cdots & \\ b_{n1} & b_{n2} & \cdots & b_{nm} \end{bmatrix} = \begin{bmatrix} c_1, c_2, c_3, \ldots, c_m \end{bmatrix}.$$

Let us define

$$\text{Det}(A \cdot B) = \sum_s (A_s B_s) = \sum_s \det(A_s) \det(B_s).$$

We wish to prove that $\text{Det}(A \cdot B) = \det(A \cdot B)$. Observe that both $\text{Det}(A \cdot B)$ and $\det(A \cdot B)$ are linear in the row vectors $r_1, \ldots, r_m$ and column vectors $c_1, \ldots, c_m$, therefore it suffices to prove the theorem

---

[3]Cauchy, A. French mathematician, engineer and physicist. 1789–1857.
[4]Binet, J. French mathematician. 1786–1856.

only for the case that all the row vectors and column vectors of A, B are vectors of the form with 1 at one spot, and 0 at the remaining spots. It is easy to see that there are only m 1's for A and B. The 1 at the i_k spot of the ith row vector of A must match the spot of 1 in some jth column vector of B. Otherwise, we will have a 0 for both $\text{Det}(A \cdot B)$ and $\det(A \cdot B)$, and they are equal. We assume that there is a 1–1 match from the 1's on the rows $r_1, \ldots, r_m$ of A to some column $c_1, \ldots, c_m$ of B. Let the 1 on the ith row of A at (ii_k) and on the matching i_j the column of B at $(i_k i_j)$. We will conclude that either (1) there are m distinct i_k's hence they form a set s of m elements, and in this case $\text{Det}(A \cdot B) = \det A_s \det B_s = \det(A \cdot B)$. Or (2) both sides are zeroes. The detailed arguments are left to the reader as an exercise. ∎

Recall the following definition in $\mathbb{R}^3$ about *cross product*,

Example 3.5 (Cross Product in $\mathbb{R}^3$): Let $v = [a_{21}, a_{22}, a_{23}]^T$ and $u = [a_{31}, a_{32}, a_{33}]^T$, then we define the cross product $v \times u$ as

$$v \times u = \det(A) = -\det \begin{bmatrix} \mathbf{i} & \mathbf{j} & \mathbf{k} \\ a_{21} & a_{22} & a_{23} \\ a_{31} & a_{32} & a_{33} \end{bmatrix}$$

$$= (a_{22}a_{33} - a_{23}a_{32})\mathbf{i} - (a_{21}a_{33} - a_{23}a_{31})\mathbf{j} + (a_{21}a_{32} - a_{22}a_{31})\mathbf{k}.$$

∎

Example 3.6: It is interesting to study the area function $\mathbb{A}$ in $\mathbb{R}^3$. Let $v = [v_1, v_2, v_3]^T$ and $u = [u_1, u_2, u_3]^T$ be two vectors in $\mathbb{R}^3$. Then the square of the area $\mathbb{A}$ is the product of the transpose of the determinant of matrix $B = [v, u]$ with itself, i.e.,

$$\mathbb{A}^2 = \det B^T B.$$

It follows from the Cauchy–Binet theorem that

$$\det B^T B = \left(\det \begin{bmatrix} v_2 & v_3 \\ u_2 & u_3 \end{bmatrix} \right)^2 + \left(\det \begin{bmatrix} v_3 & v_1 \\ u_3 & u_1 \end{bmatrix} \right)^2$$

$$+ \left(\det \begin{bmatrix} v_1 & v_2 \\ u_1 & u_2 \end{bmatrix} \right)^2.$$

From the above, it is easy to see that this definition of area $\mathbb{A}$ agrees with the definition of area vector using *cross product*. We have,

$$\mathbb{A} = v \times u = \det \begin{bmatrix} i & j & k \\ v_1 & v_2 & v_3 \\ u_1 & u_2 & u_3 \end{bmatrix}$$

$$= \det \begin{bmatrix} v_2 & v_3 \\ u_2 & u_3 \end{bmatrix} i + \det \begin{bmatrix} v_3 & v_1 \\ u_3 & u_1 \end{bmatrix} j + \det \begin{bmatrix} v_1 & v_2 \\ u_1 & u_2 \end{bmatrix} k. \qquad \blacksquare$$

Exercises

(1) Let u_1, u_2, v_1, v_2 be four vectors in $\mathbb{R}^3$. Show that

$$\begin{bmatrix} v_1^T \\ v_2^T \end{bmatrix} \cdot [u_1 \quad u_2] = \begin{bmatrix} v_1^T u_1 & v_1^T u_2 \\ v_2^T u_1 & v_2^T u_2 \end{bmatrix} = (u_1 \times u_2) \cdot (v_1 \times v_2).$$

(2) Let $u_1, u_2, u_3, v_1, v_2, v_3$ be six vectors in $\mathbb{R}^3$. Show that

$$(v_1 \cdot (v_2 \times v_3)) = \det \begin{bmatrix} v_1^T \\ v_2^T \\ v_3^T \end{bmatrix}.$$

Further show that

$$\det \begin{bmatrix} v_1^T u_1 & v_1^T u_2 & v_1^T u_3 \\ v_2^T u_1 & v_2^T u_2 & v_2^T u_3 \\ v_3^T u_1 & v_3^T u_2 & v_3^T u_3 \end{bmatrix} = (v_1 \cdot (v_2 \times v_3))(u_1 \cdot (u_2 \times u_3)).$$

(3) Find the area of the parallelogram in the real 4-dimensional space $\mathbb{R}^4$ spanned by

$$\begin{bmatrix} 0 \\ 1 \\ 2 \\ 3 \end{bmatrix} \quad \text{and} \quad \begin{bmatrix} 1 \\ 2 \\ 3 \\ 4 \end{bmatrix}.$$

(4) Let A and B be the following matrices,

$$A = \begin{bmatrix} 0 & 2 \\ 1 & 3 \\ 2 & 4 \\ 3 & 5 \end{bmatrix} \quad \text{and} \quad B = \begin{bmatrix} 1 & 4 \\ 2 & 3 \\ 3 & 2 \\ 4 & 1 \end{bmatrix}.$$

Use the Cauchy–Binet theorem to find $\det(A^T B)$.

(5) Let A and B be the following matrices,

$$A = \begin{bmatrix} 0 & 2 & 0 \\ 1 & 3 & 0 \\ 2 & 4 & 1 \\ 3 & 5 & 1 \end{bmatrix} \quad \text{and} \quad B = \begin{bmatrix} 1 & 4 & 0 \\ 2 & 3 & 1 \\ 3 & 2 & 1 \\ 4 & 1 & 0 \end{bmatrix}.$$

Use the Cauchy–Binet theorem to find $\det(A^T B)$.

(6) Finish the proof of Theorem 3.2.2.

3.3 Tensor Product and Exterior Product

In Section 1, the definition of determinant by axioms of properties is useful. We shall imitate it and define many concepts by abstract properties: the universal properties.

The concept of tensor was developed by Gauss, Riemann,[5] etc. for geometry to compare data at different positions. It was known as *tensor analysis* which was populated by the *relativity* of A. Einstein.[6] The modern and abstract treatment of tensor analysis involves the *tensor product* of modules.

From another point of view, according to the axiomatic definition of determinant, it is a *multilinear map*. Let U, V be two modules over a ring R and $U \times V$ be the module of all finite sums $\sum a_i u_i \times v_i$, and $\phi : U \times V \mapsto R$ a multilinear map. One way to study ϕ is to form a *tensor product*, $U \otimes V$, as in the definition below.

---

[5]Riemann, G. F. B. German mathematician who contributed to algebraic geometry, differential geometry, analysis and number theory. 1826–1866.

[6]Einstein, A. German-born, citizen of several countries. He made fundamental contributions to *special relativity, general relativity, quantum theory* and *Brownian motions*. In 1921, he received the Nobel Prize in Physics. 1879–1955.

Definition (Tensor Product): Let U, V be two modules over R. Let I be the submodule of $U \times V$ generated by elements of the form

$$(u_1 + u_2, v) - (u_1, v) - (u_2, v), \qquad (u, v_1 + v_2) - (u, v_1) - (u, v_2),$$

$$(ru, v) - (u, rv).$$

We define the tensor product of U, V over the ring R as $M \otimes_R N = (U \times V)/I$.

Then $U \otimes_R V$ is an R-module by defining

$$r \left(\sum_{finite} u_i \otimes v_i \right) = \sum_{finite} r u_i \otimes v_i. \qquad \blacksquare$$

As in the discussions of the definitions of determinants, we shall give a definition of tensor product abstractly.

Proposition 3.3.1: *Let U, V be R-modules. Then an R-module W is isomorphic to $U \otimes_R V$ as a commutative additive group iff there is a map $\phi : U \times V \mapsto W$ such that for any R-module G and bilinear homomorphism σ in the following diagram, we can find the map $\bar{\sigma} : W \mapsto G$ with the diagram commutes,*

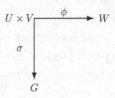

As in the following diagram,

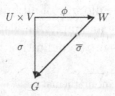

Proof. Let $G = U \otimes_R V$ and σ the homomorphism which sends (u, v) to $u \otimes v$. Then we have $\bar{\sigma} : W \mapsto U \otimes_R V$. On the other hand,

$U \otimes_R V$ has the property of W. We interchange W and $U \otimes_R V$. Then we will get a homomorphism $\psi : U \otimes_R V \mapsto W$. It is easy to see that $\overline{\sigma}\psi = identity$ and $\psi\overline{\sigma} = identity$. Therefore W and $U \otimes_R V$ are isomorphic. ∎

The above-mentioned property is significant. As usual, a good proposition turns to a good definition. We have,

Definition (Universal Property): Let U, V be modules. Then a module W is defined to be $U \otimes_R V$ as commutative additive group if there is a map $\phi : U \times V \mapsto W$ such that for any R-module G and bilinear homomorphism σ in the following diagram, we can find the map $\overline{\sigma} : W \mapsto G$ with the diagram commutes,

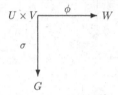

There is a homomorphism $\overline{\sigma}$ to complete the above diagram.

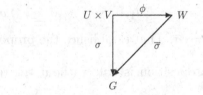

We have the following useful corollary,

Corollary 3.3.2: *Let the two fields $L \supset_R R$. Then $L \otimes_R R \approx L$.*

Proof. We define $\sigma(\ell, k) = \ell k$. It is routine to see that $\overline{\sigma}$ is an isomorphism. ∎

We have the following proposition about the commutative property between the tensor product and direct sum,

Proposition 3.3.3: *Given R-modules $\{U_i\}$ and $\{V_j\}$, we have*

$$\left(\sum_{finite} U_i\right) \otimes_R V_i \approx \sum_{finite} (U_j \otimes_R V_i).$$

Proof. It suffice to show

$$(U_1 \oplus U_2) \otimes_R V_1 \approx U_1 \otimes_R V_1 \oplus U_2 \otimes_R V_1.$$

The general case is similar.

We simply show that $U_1 \otimes V_1 \oplus U_2 \otimes V_1$ has the universal property of the tensor product of $(U_1 \oplus U_2)$ and V_1 in the following diagram,

$$\begin{array}{ccc}
(U_1 \oplus U_2) \times V_1 & \xrightarrow{\phi} & (U_1 \oplus U_2) \otimes_R V_1 \\
{\scriptstyle \sigma}\downarrow & {\scriptstyle \overline{\sigma}} & \\
(U_1 \otimes_R V_1) \oplus (U_2 \otimes_R V_1) & &
\end{array}$$

The map $\overline{\sigma}$ is defined by $\overline{\sigma}((u_1 + u_2) \otimes v_1) = (u_1 \otimes v_1, u_2 \otimes v_1)$. It is well-defined, since u_1, u_2 are uniquely determined. On the other hand, a map τ defined by,

$$\tau(u_1 \otimes v_1, u_2 \otimes v_1) = (u_1, 0) \otimes v_1 + (0, u_2) \otimes v_1$$

clearly satisfies $\overline{\sigma}\tau = 1 = \tau\overline{\sigma}$. That establishes the proposition. ∎

Remark: The above proposition is true without the restriction of *finite*. ∎

We have the following useful corollary,

Corollary 3.3.4: *Let the two fields $L \supset K$. Then $L \otimes_K K^n \approx L^n$.* ∎

Example 3.7 (Extension of Scalar): Let S be an overfield of K and U be a vector space over K. Then $S \otimes_K U$ is naturally a vector space over S by defining

$$s\left(\sum_i s_i \otimes u_i\right) = \left(\sum_i (ss_i) \otimes u_i\right).$$

For instance, let the complex numbers $\mathbb{C} \supset \mathbb{R}$ be the real numbers and let V be a real vector space, we take $\mathbb{C} \otimes_\mathbb{R} V$ as a complex vector space. In other words, we extend the scalar of real numbers $\mathbb{R}$ to complex numbers $\mathbb{C}$. We have $\mathbb{R} \otimes_\mathbb{R} V = V$. ∎

Example 3.8: Let us consider $\mathbb{C} \otimes_\mathbb{R} \mathbb{C}$ where $\mathbb{C}$ is the field of complex numbers and $\mathbb{R}$ is the real number field. The complex number field $\mathbb{C}$ is viewed as a vector space with basis $\{1, i\}$. The scalar extension will make $\mathbb{C} \otimes_\mathbb{R} \mathbb{C}$ as a complex vector space of dimension 2 over $\mathbb{C}$ while as a real vector space of dimension 4. Its real bases are $e_1 = 1 \otimes 1, e_2 = i \otimes 1, e_3 = 1 \otimes i, e_4 = i \otimes i$. We have $e_4^2 = i^2 \otimes i^2 = -1 \otimes -1 = 1$. In other words, $(e_4^2 - 1) = (e_4 - 1)(e_4 + 1) = 0$. Therefore $\mathbb{C} \otimes_\mathbb{R} \mathbb{C}$ is not an integral domain, certainly not a field. ∎

Definition (Remark): It is easy to see that the above definition can be easily extended to three or more components of R-modules. We have $(U \otimes_V) \otimes_R W \approx U \otimes (V \otimes_R W)$. We may write it as $U \otimes_R V \otimes_R W$. We may further generalize to any finite number of components in the tensor product. Especially, we have,

$$T^i(U) = \overbrace{U \otimes_R U \otimes_R \cdots \otimes_R U}^{i}$$

and set $T^0(U) = R$, $T(U) = \oplus_{i=0}^\infty T^i(U)$. In this way, $T(U)$ is an *algebra*, the *tensor algebra* over U. ∎

We define

Definition (Exterior Algebra): The ith *exterior module*, $\Lambda^i(U)$, is defined to be $T^i(U)/A^i(U)$ where $A^i(U)$ is the submodule of $T^i(U)$ generated by all elements of the form $u \otimes u$. The image of $u_1 \otimes u_2 \otimes \cdots u_i$ is written as $u_1 \wedge u_2 \wedge \cdots \wedge u_i$. We define the *exterior algebra*, $\Lambda(U)$, as $\oplus_{i=0}^\infty \Lambda^i(U)$. It is an R-algebra. ∎

We have the following proposition which defines the exterior product by the *universal property*:

Proposition 3.3.5: *Given a module U. For some module W, and a map ϕ from $U \times U$ to W such that for any alternative and bilinear map f from module $U \times U$ to R, then there is a linear map $\overline{f}$ from W to R such that the following diagram commutes $\Leftrightarrow$ W is the exterior product of $\Lambda^2 U$.*

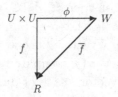

Proof. It is left to the readers. ∎

The above proposition is the usual *universal property* of an exterior product. It is similar to the axiomatic definition of the determinant.

Example 3.9 (Cross Product in $\mathbb{R}^3$ and Exterior Product):

Let $v = [a_{21}, a_{22}, a_{23}]^T$ and $u = [a_{31}, a_{32}, a_{33}]^T$. The definition of the *cross product* $v \times u$ in vector analysis is

$$v \times u = \det(A) = -\det \begin{bmatrix} \mathbf{i} & \mathbf{j} & \mathbf{k} \\ a_{21} & a_{22} & a_{23} \\ a_{31} & a_{32} & a_{33} \end{bmatrix}$$

$$= (a_{22}a_{33} - a_{23}a_{32})\mathbf{i} - (a_{21}a_{33} - a_{23}a_{31})\mathbf{j} + (a_{21}a_{32} - a_{22}a_{31})\mathbf{k}.$$

On the other hand, we have

$$v \wedge u = (a_{21}\mathbf{i} + a_{22}\mathbf{j} + a_{23}\mathbf{k}) \wedge (a_{31}\mathbf{i} + a_{32}\mathbf{j} + a_{33}\mathbf{k})$$

$$= (a_{22}a_{33} - a_{23}a_{32})\mathbf{j} \wedge \mathbf{k} - (a_{21}a_{33} - a_{23}a_{31})\mathbf{k} \wedge \mathbf{i}$$

$$+ (a_{21}a_{32} - a_{22}a_{31})\mathbf{i} \wedge \mathbf{j}.$$

It is clear that the only difference between the *cross product* and the *exterior product* is the notations of the basis of the area. For the basis of area, $\mathbf{i} \wedge \mathbf{j}$, $\mathbf{j} \wedge \mathbf{k}$, $\mathbf{k} \wedge \mathbf{i}$, we substitute them by their normal directions $\mathbf{k}, \mathbf{i}, \mathbf{j}$ (by Hodge star actions), then we get the ordinary cross product.

Only in $\mathbb{R}^3$, we have the cross product of vectors, while in any dimensional real spaces we have the exterior product of any degree.

In $\mathbb{R}^4$, we have the 6-dimensional $e_i \wedge e_j$ with $i \neq j$, as $\sum_{i<j} a_{ij} e_i \wedge e_j$ of degree 2. Each of them defines an area which is $\sqrt{\sum a_{ij}^2}$, according to the Cauchy–Binet theorem, the m-dimensional volume of the parallelopiped spanned by m vectors $v_1, v_2, \ldots, v_m$ is the absolute value of the vector $v_1 \wedge v_2 \cdots \wedge v_m$. It makes all things fit. $\blacksquare$

Example 3.10: Let V be a vector space over a field K of dimension n. Then $\Lambda^0(V) = K$, $\Lambda^1(V) = V$.

In general, $\dim \Lambda^i(V) = \binom{n}{i}$ for all $1 \leq i \leq n$ and $\Lambda^{n+i}(V) = 0$ for all $i \geq 1$. Especially, $\dim \Lambda^n(V) = 1$. Let $e_1, e_2, \ldots, e_n$ be the standard basis of K^n. Then the determinant can be viewed as a linear map of $\Lambda^n(K^n) \mapsto K$ which sends $e_1 \wedge e_2 \wedge \cdots \wedge e_n \mapsto 1$. Note that this definition of determinant is not coordinate free. $\blacksquare$

Definition Let us consider an orthonormal basis $\{e_1, e_2, \ldots, e_n\}$ of $\mathbb{R}^n$. Let us consider m vectors $v_1, \ldots, v_m$ where

$$v_i = \sum a_{ij} e_j.$$

We consider the exterior product $v_1 \wedge \cdots \wedge v_m$ and the matrix A whose column vectors are $[a_{i1}, \ldots, a_{in}]^T$. Then there are $= \binom{n}{m}$ many subsets of n with precise m elements. Let the set of subsets of n with m elements be S. Let $e_s = e_{i_1}, \ldots, e_{i_m}$ where $s = \{i_1, \ldots, i_m\}$ in a fixed order, and here A_s is the $m \times m$ matrix that consists of the m columns determined by s. Then we have that

$$v_1 \wedge \cdots \wedge v_m = \sum_s \det(A_s) e_s.$$

We shall define it as *the vector of m-dimensional volume* $|v_1, \ldots, v_m|$ of the vectors $v_1, v_2, \ldots, v_m$. According to the previous section, we have

$$|v_1, \ldots, v_m| = \sqrt{\det(A^T A)} = \mathbb{A}. \qquad \blacksquare$$

Example 3.11 (Jacobian Determinant in Integration): Let us consider the integration on the plane,

$$\int_\Omega f(x, y) dx \wedge dy.$$

Sometimes it is convenient to change the variable from (x, y) to (u, v). Let $x = x(u, v)$, $y = y(u, v)$. Then we have $dx = \frac{\partial x}{\partial u} du + \frac{\partial x}{\partial v} dv$, and $dy = \frac{\partial y}{\partial u} du + \frac{\partial y}{\partial v} dv$. By direct computation, we have $dx \wedge dy = J(du \wedge dv)$ where

$$J = \det \begin{bmatrix} \dfrac{\partial x}{\partial u} & \dfrac{\partial x}{\partial v} \\[2ex] \dfrac{\partial y}{\partial u} & \dfrac{\partial y}{\partial v} \end{bmatrix}$$

and the integration gets transferred to

$$\int_{\Omega^*} Jf(x(u, v), y(u, v)) du \wedge dv.$$

Similar changes will show up in n-variable cases. ∎

Exercises

(1) Show that there is a unique R-vector space linear map $U \otimes_R V \cong V \otimes_R U$ which sends $u \otimes v$ to $v \otimes u$.

(2) Show that there are unique K-vector space isomorphisms $(U_1 \oplus U_2) \otimes_R V \cong (U_1 \otimes_R V) \oplus (U_2 \otimes_R V)$ and $U \otimes_R (V_1 \oplus V_2) \cong (U \otimes_R V_1) \oplus (U \otimes_R V_2)$ such that $(u_1, u_2) \otimes v \mapsto (u_1 \otimes v, u_2 \otimes v)$ and $u \otimes (v_1, v_2) \mapsto (u \otimes v_1, u \otimes v_2)$, respectively.

(3) Prove that if ϕ is an alternating bilinear form on M, then ϕ is anti-symmetric, i.e., $\phi(x, y) = -\phi(y, x)$.

(4) Suppose that $2 \neq 0$. Then every anti-symmetric linear form is alternating.

(5) Suppose that $2 = 0$. Find an example that is an anti-symmetric bilinear form which is not alternating.

(6) Prove Proposition 3.3.5.

(7) Prove Proposition 3.3.6.

3.4   Dual Space. Application to Physics

The whole theory of *Calculus* is built on the concepts of *functions* and *limits*. Usually a function is defined on a subset of $\mathbb{R}$, the domain of f, with range $\mathbb{R}$. For vector space V, we require the domain to be the vector space V, and the range to be the field K. In this section, we will restrict our attention to the simplest functions, which are homogeneous and linear. Furthermore we will not study functions as individual entities, we will rather look at them as a space. We define,

Definition Given a module U over a ring R, the collection of all homomorphic functions from R-module U to R-module R will form a module, $\text{Hom}_R(U, R) = U^*$, over R. This module is called the *dual module* of U. The elements in $\text{Hom}_K(U, K)$ are called *linear functionals* or simply *functionals*. ∎

Remark: If the ring R is a field K, it is easy to show that $\dim(U) = \dim(U^*)$ (or see below), if U is finite dimensional. If U is infinite dimensional, then we may have $\dim(U^*) > \dim(U)$. We have examples to show that it is possible to have $\dim(U^{**}) > \dim(U^*) > \dim(U)$. (See the Exercises.) ∎

Proposition 3.4.1: *We have* $\text{Hom}_R(R, R) = R$.

Proof. Let $f \in \text{Hom}_R(R, R)$. Then $f(r) = f(r.1) = r \cdot f(1)$, $f(1) \in R$, is determined by and will determine f. ∎

Let $\{v_1, v_2, \ldots, v_n\}$ be a basis of a finitely dimensional vector space U over K. Then the basis induces a dot product pairing ∘ for U. As for $v = \sum a_i v_i$, $u = \sum b_i v_i$, we define $v \circ u = \sum a_i b_i$. Note that $v_i \circ v_j = \delta_{ij}$.

Proposition 3.4.2: *Let U be a finitely dimensional K-vector space. Then $\text{Hom}_K(U, K)$ consists of all functionals defined by a dot product with a fixed element $v \in U$, i.e., all f_v such that $f_v(u) = v \circ u$. Thus $f : U \cong \text{Hom}_K(U, K)$ by $f(v) = f_v$. Especially, $\dim(U) = \dim(U^*)$.*

Proof. Clearly a functional f_v defined by the dot product with a fixed element v is in $\text{Hom}_K(U, K)$, therefore the map f which sends v to f_v defines $U \mapsto \text{Hom}_K(U, K)$. We wish to show it is 1–1 and onto.

Let $e_1, e_2, \ldots, e_n$ be the basis which defines the dot product. If $f_v = 0$ and $v = a_1 e_1 + a_2 e_2 + \cdots + a_n e_n$, then $f_v(e_i) = a_i = 0$ for all i. Henceforth $v = 0$. So it is 1–1. On the other hand, let $u = b_1 e_1 + b_2 e_2 + \cdots + b_n e_n$ and g is any linear functional with $a_i = g(e_i)$ for $i = 1, \ldots, n$, then we have $g(u) = (b_1 g(e_1) + b_2 g(e_2) + \cdots + b_n g(e_n)) = (b_1, b_2, \ldots, b_n) \cdot (g(e_1), g(e_2), \ldots, g(e_n))^T = (b_1, b_2, \ldots, b_n) \circ (a_1, a_2, \ldots, a_n)^T$. We let $v = a_1 e_1 + a_2 e_2 + \cdots + a_n e_n$, then $g = f_v$ and it is onto. ∎

In general we have the following propositions,

Proposition 3.4.3: *Let M, N be two R-modules. Then $(M \oplus N)^* \cong M^* \oplus N^*$.*

Proof. Let $\phi \in (M \oplus N)^*$. Then define $\phi_1(m) = \phi(m, 0)$ and $\phi_2(n) = \phi(0, n)$. Clearly $\phi_1 \in M^*, \phi_2 \in N^*$. On the other hand, let $\varphi_1 \in M^*$, $\varphi_2 \in N^*$. Define $\varphi(m, n) = \varphi_2(m) + \varphi_2(n)$, then $\varphi \in (M \oplus N)^*$. ∎

Proposition 3.4.4: *Let M_i $(i \in I)$ be a family of R-modules. Then $(\oplus_i M_i)^* \cong \prod_{i \in I} M_i^*$.*

Proof. Notice that M_i can be embedded to $(\oplus_i M_i)$ as we take all other components 0. We define a map Φ

$$\Phi : (\oplus_i M_i)^* \mapsto \prod_{i \in I} M_i^*$$

with any $\psi \in (\oplus_i M_i)^*$, i.e., $\psi_i = \psi|_{M_i} \in M_i^*$, $\Phi(\psi) = \prod_i \psi_i \in \prod_i M_i^*$. Its inverse map Φ^{-1} is the following map: given any element $\prod_i \varphi_i \in \prod_i M_i^*$. $\Phi^{-1}(\prod_i \varphi_i) = \psi$, where $\psi(\oplus_i m_i) = \sum_i \varphi_i(m_i)$ for $\oplus_i m_i \in \oplus_i M_i$. Note that in $\oplus_i M_i$, there are only finitely many non-zero terms. ∎

Remark: The same proof shows that

$$\mathrm{Hom}_R(\oplus_{i \in I} M_i, N) \cong \prod_{i \in I} \mathrm{Hom}_R(M_i, N)$$

where the notation $\mathrm{Hom}_R(M, N)$ is the R-module of all homomorphisms of R-module M to R-module N. ∎

Example 3.12: Let $M = \oplus_{i \geq 0} \mathbb{Z}$ the direct sum of countable copies of $\mathbb{Z}$. Then M^* is not free. (See the Exercises.) ∎

Definition Let $M = R \oplus R \oplus \cdots \oplus R$ be a finite free R-module. Let $e_1, e_2, \ldots, e_n$ be a basis of M. Then $\{e_i^* : 1 \leq i \leq n\}$ defined by

$$e_i^* \left(\sum_j a_j e_j \right) = a_i$$

is called dual basis of M^*. ∎

Example 3.13: Let $\mathbb{R}$ be the field of real numbers and $M = \mathbb{C}$ the field of complex numbers considered to be a module over $\mathbb{R}$. We take a basis $\{1, i\}$ for M over $\mathbb{R}$. Then the dual basis is $\{1^*, i^*\}$ where

$$1^*(a + bi) = a, \quad i^*(a + bi) = b.$$

They are just the real part function and the imaginary part function. ∎

Proposition 3.4.5: *Let M be a finite free module over a ring R. Then there is a natural map (i.e., coordinate free map) $\phi : M \approx M^{**}$.*

Proof. We shall define a map from M to M^{**} and then show it is an isomorphism. Let $m \in M$, $\phi \in M^*$. We shall write the function $\phi(m)$ as $\langle \phi, m \rangle$. We may fix m and vary ϕ. Then clearly m defined one element in M^{**}. Let us call it ev_m. Our map will send m to ev_m.

Now we will restrict our attention to the case of finite free modules. In this case we may pick a basis $e_1, \ldots, e_n$, and let the dual basis be $e_1^*, \ldots, e_n^*$. First we will show that if $m \neq 0$, then $ev_m \neq 0$, i.e., let $m = \sum_i a_i e_i$ with some $a_i \neq 0$, then $ev_m(e_i^*) = a_i \neq 0$. Hence $ev_m \neq 0$. In other words, the correspondence is one-to-one.

Now we shall show the map m to ev_m is onto. Let $\psi \in M^{**}$ and $\psi(e_i^*) = a_i$. Let $m = \sum_i a_i e_i$. Then it is easy to check that $ev_m(e_i^*) = a_i$. Hence $\psi = ev_m$. ∎

Proposition 3.4.6: *Let R be an integral domain with K its field of fractions. Let I be any ideal of R. Then $I^* \approx [R : I] = \{c \in K : cI \subset R\}$.*

Proof. We claim that for any element $\phi \in I^*$, it corresponds to an element in K in the following way, let $\phi(a) = b$, $\phi(c) = d$ where $a, c \in I$, and $b, d \in R$. Then $\phi(ac) = cb = ad$. Therefore $b/a = d/c = \alpha \in K$. It is easy to see that the effect of the map ϕ is the same as multiplying by α. ∎

Example 3.14: We have to evaluate many items, a group of journals, or a group of cars, etc., for our selection. There are several criteria for all items in a group for us to pass judgments on. Usually, we depend on the judgments of experts on the relative importance of the criteria. We may put everything in terms of mathematics as follows. Let each item v of the group to be evaluated be a vector in $\mathbb{R}^n$ spanned by the basis $\{e_1, \ldots, e_n\}$, where each coordinate represents a criteria. The value scored by v with respect to e_j is the jth coordinate v_j. The weight m_j on criteria e_j assigned by the experts form a linear functional $\sigma = \sum_j m_j e_j^*$ of $(\mathbb{R}^n)^*$. The value of v is $\sigma(v)$.

The above is the most popular way of using linear functional to evaluate a group of multicriteria items. In the later Section 6.2, we will present the non-linear way of evaluations. ∎

Example 3.15: We will give an example that the two modules M and M^{**} are not isomorphic. Let $R = \mathbb{Z}[x]$ and $M = (2, x)$ be the ideals generated by 2 and x. From the previous proposition, we know that $M^* \approx [R : I] = \{c \in K : cI \subset R\}$ where $K = \mathbb{Q}(x)$. Let $c = f(x)/g(x)$. Then it is easy to see that $2c \in R$ and $xc \in R$ implies $c \in R$. We conclude that $[R : I] = R$ and $M^{**} = R^* = R$. Therefore M is not isomorphic to M^{**}. ∎

Example 3.16: We will give an example that the two modules M and M^{**} are isomorphic while M is not a finite free R-module. Let $R = \mathbb{Z}[\sqrt{-14}]$, $M = I = (3, 1 + \sqrt{-14})$, $J = (3, 1 - \sqrt{-14})$. Let us consider IJ. Clearly $9 = 3 \cdot 3 \in IJ$ and $6 = 3 \cdot (1 + \sqrt{-14} + 1 - \sqrt{-14}) \in IJ$, $3 = 9 - 6 \in IJ$. Then it is easy to see that $IJ = (3)$. We claim that I is not principal. Assume the other way. Let $I = (\alpha)$ with $\alpha = a + b\sqrt{-14}$. Then we have the following equation,

$$3 = \beta\alpha = \beta(a + b\sqrt{-14}).$$

If we consider the complex *norm* on both sides, we have

$$9 \geq |\alpha| = a^2 + b^2 14$$

which is possible only if $b = 0$, then we have $9 = |\beta||\alpha|$, and therefore $a = \pm 1, \pm 3$. Let us consider the case $\alpha = 3$ (similarly $\alpha = -3$). Then we have

$$1 + \sqrt{-14} = (c + d\sqrt{-14})3$$

which produces the following equations,

$$1 = c3, \quad 1 = 3d.$$

It is clearly impossible in $\mathbb{Z}$. Similar arguments work for $a = -3$. Let us consider the case $\alpha = 1$. Note that $(\alpha) = I = (3, 1 + \sqrt{-14})$, then we have

$$1 = 3(r + s\sqrt{-14}) + (u + v\sqrt{-14})(1 + \sqrt{-14}).$$

We shall expand and compare the coefficients of 1 and $\sqrt{-14}$; we get

$$u + v = -3s, \quad 3r + u - 14v = 1, \quad \text{or} \quad 3s - 3r - 15v = 1.$$

The last equation is quite impossible. Similarly, $\alpha \neq -1$.

So we conclude that the ideal I is not principal, i.e., they are not free. It is not hard to see that $I^* = [R : I] = (1/3)J \approx J$. By a similar argument, we see that $J^* \approx I$. It follows that $I^{**} \approx I$, $J^{**} \approx J$, and I, J are both not free. ∎

Naturally, we will generalize the dual module, $M^*(= \mathrm{Hom}_R(M, R))$ to $\mathrm{Hom}_R(M, N)$, the set of all R-homomorphisms of M to N. Let $\phi \in \mathrm{Hom}_R(M, N)$ and $\sigma \in \mathrm{Hom}_R(N, R)$. We have the following definition,

Definition We define ϕ^*, the dual of a map $\phi \in \mathrm{Hom}_R(M, N)$. In the following diagram,

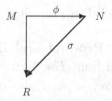

Let us define $\phi^*(\sigma) = \sigma \cdot \phi$. Then $\phi^* : \mathrm{Hom}_R(N, R) \mapsto \mathrm{Hom}_R(M, R)$. ∎

Let $\phi \in \operatorname{Hom}_R(M, N)$, M, N be finite free modules and $\{e_1, e_2, \ldots, e_n\}$ be a basis of M, $\{d_1, d_2, \ldots, d_m\}$ be a basis of N. Let

$$\phi(e_i) = \sum_j a_{ji} d_j.$$

We may use the notation of matrices. Let us write $u = \sum_i a_i e_i$ as $(a_1, \ldots, a_m)^T$, and $v = \sum_j b_j d_j$ as $(b_1, \ldots, b_n)^T$. Let $\{e_1^*, e_2^*, \ldots, e_n^*\}$ be the dual basis of $\operatorname{Hom}_R(M, R)$, and $\{d_1^*, d_2^*, \ldots, d_m^*\}$ be the dual basis of $\operatorname{Hom}_R(N, R)$. Let us write $u^* = \sum_i c_i e_i^*$ as $(c_1, \ldots, c_m)$. Let $\{d_1^*, d_2^*, \ldots, d_m^*\}$ be the dual basis of $\operatorname{Hom}_R(N, R)$. Let us write $v^* = \sum_j p_j d_j^*$ as $(p_1, \ldots, p_n)$. Let $A = (a_{ij})$. Then we always have $\phi(u) = v \iff (b_1, \ldots, b_n)^T = A(a_1, \ldots, a_m)^T$. We have the following proposition for ϕ^*.

Proposition 3.4.7: *Let us use the notations of the preceding paragraph. Then we have $\phi^*(v^*) = u^* \iff (p_1, \ldots, p_n)A^T = (c_1, \ldots, c_m)$. Note that the matrix of ϕ^* with respect to those dual bases is A^T.*

Proof. Let us compute $\phi^*(d_k^*)$. We have $\phi^*(d_k^*)(e_i) = d_k^*\phi(e_i) = d_k^* \sum_j a_{ji} d_j = a_{ki}$. Therefore $\phi^*(d_k^*) = \sum_i a_{ki} d_i^*$. The proposition follows. ∎

Example 3.17: Let W, V be finite dimensional vector spaces of dimension m, n respectively. Let us consider $W^* \otimes V$. We define $(f \otimes v)(w) = f(w)v$ and generalize the definition to $W^* \otimes V$. Let $e_1, \ldots, e_m$ be a basis of W, and $v_1, \ldots, v_n$ be a basis for V. Let $f_1, \ldots, f_m$ be a dual basis for W^*, i.e., $f_i(e_j) = \delta_{ij}$. Then it is easy to see that $f_i \otimes v_j$ forms a basis for $W^* \otimes V$. It is routine to show that $W^* \otimes V$ is isomorphic to $M_{n \times m} - \{all \ n \times m \ matrices\}$. ∎

Example 3.18 (Tensor Product and Dual in Physics): The following example is taken from *The Feynman*[7] *Lectures on Physics*, p. 31-1, Vol II. We are given a particular crystalline substance. Let

---

[7]Feynman, R. American physicist. 1965 Nobel Prize in Physics. 1918–1988.

$E = (E_1, E_2, E_3)^T$ be the electric field (which is a vector field), and $P = (P_1, P_2, P_3)^T$ be the polarization. Then their relations can be expressed as

$$\begin{bmatrix} P_1 = \alpha_{11}E_1 + \alpha_{12}E_2 + \alpha_{13}E_3 \\ P_2 = \alpha_{21}E_1 + \alpha_{22}E_2 + \alpha_{23}E_3 \\ P_3 = \alpha_{31}E_1 + \alpha_{32}E_2 + \alpha_{33}E_3 \end{bmatrix}.$$

We may write the above as $P = AE$, where $A = (\alpha_{ij})$. Then the matrix (α_{ij}) is called the matrix of *polarization*. Note that $\mu_P = (1/2)E^T P$ is the work done to bring the polarization from 0 to P. The work is a scalar independent of the coordinate system.

Now we change to a new coordinate system x_1', x_2', x_3'. Let $E' = BE$, $P' = CP$. Since $E^T P$ is a scalar, therefore we have $B^T C = 1$, or $C = (B^T)^{-1}$. Let $P' = A'E'$. Then we have $A = B^T A'B$.

Let $S =$ all infinite differential functions in three variables, and $K_1 = SE_1 \oplus SE_2 \oplus SE_3$. Let K_1^* be the dual module of K_1. We conclude that the matrix $A = (\alpha_{ij})$ is transformed as an element in $K_1 \otimes K_1^*$, we may thus identify it with $\sum_{ij} \alpha_{ij} E_i \otimes p_j$ where p_j is in the dual space K_1^*. ∎

Exercises

(1) Find a vector space U over a field K such that $\dim(U^{**}) > \dim(U^*) > \dim(U)$.

(2) Show that countable direct product of $\mathbb{Z}$ is not free.

(3) Finish the discussion of Example 3.12.

(4) Let V be a subspace of dimension $n - 1$ of an n-dimensional vector space U. Let ϕ be an endomorphism of U (i.e., $\phi(U) \subset U$) which fixes every element in V. We know that there is an element α such that $\phi(u) = au + t_u$, where a is common for every u and $t_u \in V$. If $a = 1$, then let $g(x)$ be a non-zero functional such that $g(V) = 0$, show that there is a unique element v such that $\phi(u) = u + g(u)v$.

(5) Suppose that y_1, y_2 are linear functional on K^3. Find conditions on y_1, y_2 such that for any set of scalars b_1, b_2, there is always a vector x, with $y_1(x) = b_1$ and $y_2(x) = b_2$.

(6) Let $\mathbb{C}$ be the field of complex numbers. Let $\phi : \mathbb{C} \mapsto \mathbb{C}$ be defined as $\phi(z) = (1+i)z + z$. Show that ϕ is linear, and find the matrix of ϕ^* with respect to $1, i$.

(7) Let $\phi : M \mapsto N$ be the zero map. Show that ϕ^* is the zero map.

(8) Let $\phi : M \mapsto N$ be the identity map. Show that ϕ^* is the identity map.

(9) The map $\phi \mapsto \phi^*$ is R linear.

(10) Let $\phi : M \mapsto N$ be an isomorphism of R-modules M, N. Show that ϕ^* is an isomorphism and $(\phi^*)^{-1} = (\phi^{-1})^*$.

Chapter 4

Inner Product Spaces

4.1 Inner Product

In this section we will generalize the results of Section 1.3, we assume the field is the real field $\mathbb{R}$ or complex field $\mathbb{C}$, or a subfield F of complex field $\mathbb{C}$. We mainly generalize the underlining Euclidean spaces to function spaces. We have the following definition which gives us the sense of lengths and angles.

Definition (Inner Product): Let V be a vector space over the field K where K is either the real field or the complex field. An *inner product* $\langle, \rangle$ is a function of two variables on $V \times V$ with values in K such that

(1) The function $\langle, \rangle$ is linear in the first variable.
(2) The function $\langle, \rangle$ is conjugate-symmetric, i.e., $\langle v, u \rangle = \overline{\langle u, v \rangle}$.
(3) The function $\langle, \rangle$ is positive-definite, i.e., $\langle v, v \rangle \geq 0$ for all v and $\langle v, v \rangle = 0 \iff v = 0$.

If there is an inner product on V, then we call V an *inner product space*. ∎

We use the materials in Section 1.3 to build up our sense of length and angle. From the above definition we may recover the concepts of length and angle as follows, we define the length $|v|$ of a vector as $|v|^2 = \langle v, v \rangle$, and the angle θ between two vectors v, u as $|v||u| \cos(\theta) = \mathrm{Re}(\langle v, u \rangle)$. We have to show the following proposition,

155

Proposition 4.1.1: *We have* (1) $|(\langle v, u \rangle)| \leq |v||u|$ *(it is equivalent to* $|\cos(\theta)| \leq 1$, *Cauchy–Bunyakovsky[1]–Schwarz[2] (CBS) inequality),* (2) *the triangle inequality:* $|v + u| \leq |v| + |u|$.

Proof.

(1) If $|u| = 0$, then the inequality is valid. Let us assume $|u| \neq 0$, consider the vector $re^{-i\theta}u + v$ for real numbers r, θ, where θ is defined by the polar expression of the complex number

$$\langle u, v \rangle = ke^{i\theta} \qquad k = |\langle u, v \rangle| \geq 0$$

we have $\langle re^{-i\theta}u + v, \ re^{-i\theta}u + v \rangle = |u|^2|r|^2 + r(|\langle u, v \rangle|) + r(|\langle u, v \rangle|) + |v|^2 = ar^2 + 2br + c \geq 0$ always where $a = |u|^2$, $b = |\langle u, v \rangle|$, $c = |v|^2$, which implies that the polynomial in r cannot have two distinct real roots, which further means that the discriminant $b^2 - ac \leq 0$, or $|\langle u, v \rangle| \leq |u||v|$.

(2) $|v + u|^2 = \langle v + u, v + u \rangle = |v|^2 + 2\operatorname{Re}(\langle v, u \rangle) + |u|^2 \leq |v|^2 + 2|v||u| + |u|^2 \leq (|v| + |u|)^2$ by Cauchy–Bunyakovsky–Schwarz inequality. ∎

We usually say that a triangle is strong, which means that unlike a parallelogram, a triangle cannot be deformed without changing the lengths of the three sides. Mathematically, we have the following proposition which shows that the inner product is determined by lengths,

Proposition 4.1.2: *We have the following identities,*

$$\operatorname{Re}(\langle v, u \rangle) = (1/4)(|v + u|^2 - |v - u|^2)$$

$$\operatorname{im}(\langle v, u \rangle) = (1/4)(|v - iu|^2 - |v + iu|^2).$$

Proof. It is easy. ∎

In this case after we extend the concept from dot product to inner product with suitable abstraction, we cover much more ground. We have the following example,

---

[1]Bunyakovsky, V. Russian mathematician. 1804–1889.
[2]Schwarz, H. German mathematician. 1843–1921.

Example 4.1: We may consider the space of real polynomials of degree $\leq n$, $P_n(x) = \{f(x) : f(x) \text{ polynomial, and } \deg(f) \leq n\}$. We define

$$\langle f, g \rangle = c \cdot \int_{x=a}^{x=b} f(x)g(x)dx, \quad c > 0, \ b > a.$$

It is easy to see that all axioms of the inner product are satisfied. Therefore we may discuss the length of a polynomial, and the angle between two polynomials. The lengths and angles defined in this way are not idle speculations of mathematicians, they have important applications in physics and engineering. ∎

Example 4.2: Similar to the example above, we consider all continuous functions $C[a, b]$ defined over an interval $[a, b]$ with $b > a$. We define the inner product as

$$\langle f, g \rangle = c \cdot \int_{x=a}^{x=b} f(x)g(x)dx, \quad c > 0.$$

This set of functions has a wider application in physics and engineering. ∎

Example 4.3: We claim that,

$$\left(\int_{x=a}^{x=b} f(x)g(x)dx \right)^2 \leq \left(\int_{x=a}^{x=b} f(x)^2 \, dx \right) \cdot \left(\int_{x=a}^{x=b} g(x)^2 \, dx \right).$$

Inequalities of these kinds are hard to deduce directly in *analysis*. However, it follows from the Cauchy–Bunyakovsky–Schwarz (CBS) inequality that we have $|f|^2|g|^2 \geq (\langle f, g \rangle)^2$, and the above inequality follows. ∎

Once we have the concepts of length and angle for many vector spaces, we shall define,

Definition Let V be an inner product space with inner product $\langle , \rangle$. Then a vector v is said to be a *unit* vector iff $|v| = 1$. Two vectors v, u are perpendicular to each other, symbolically $v \perp u$, iff $\langle v, u \rangle = 0$, i.e., it means that the vector spaces $\mathbb{C}v$ and $\mathbb{C}u$ are perpendicular.

Especially, the vector 0 is perpendicular to all vectors. We define the *distance* between two vectors v, u as $d(v, u) = |v - u|$. ∎

Apparently, over the complex numbers $\mathbb{C}$, the concept of two vectors v, u span a $\pi/2$ angle i.e., $\text{Re}\langle v, u \rangle = 0$, is different from the concept that they are perpendicular, i.e., $\langle v, u \rangle = 0$. For instance, two vectors $\{1, i\}$ in $\mathbb{C}$ span an angle $\pi/2$, while $\langle 1, i \rangle \neq 0$, so they are not *perpendicular*.

We could have other useful general concepts in vector spaces as,

Definition Let V be a vector space. If there is a norm $|v|$ for any vector v with (1) $|v| \geq 0$, and $|v| = 0 \Longleftrightarrow v = 0$, (2) $|av| = |a||v|$, for any $a \in R$, and (3) the triangle inequality: $|v + u| \leq |v| + |u|$. Then V is called a *normed space*. (**Remark:** A complete normed vector space is called a *Banach*[3] *space*.) ∎

We have the following,

Proposition 4.1.3: *Every inner product space is a normed space with* $|v|^2 = \langle v, v \rangle$.

Proof. The items (1) and (3) are obvious. We need only to prove the item (2) above, i.e., $|av| = |a||v|$. This follows from the definition of the inner product, $|av|^2 = \langle av, av \rangle = |a|^2 \langle v, v \rangle = |a|^2 |v|^2$. ∎

Definition Let V be a *topological* space. If for any two points v, u, there is a non-negative real-valued distance $d(v, u)$ defined with (1) $d(v, u) \geq 0$, and $d(v, u) = 0 \Longleftrightarrow v = u$, (2) $d(v, u) = d(u, v)$, and (3) the triangle inequality: $d(v, u) \leq d(v, w) + d(u, w)$, then V is called a *metric space*. ∎

Proposition 4.1.4: *Any inner product space is a metric space.*

Proof. Let $d(u, v)^2 = \langle u - v, u - v \rangle$. Then the proposition is obvious. ∎

---

[3]Banach, S. Polish mathematician. 1892–1945.

Exercises

(1) Let x, y be two vectors in an inner product space V. Show that
$$||x + y||^2 + ||x - y||^2 = 2||x||^2 + 2||y||^2.$$

(2) Let $\mathbb{R}^2$ be the real plane. Let us define
$$\langle x, y \rangle = x_1 y_1 - x_2 y_2, \quad \text{where } x = [x_1, x_2]^T, \ y = [y_1, y_2]^T.$$
Show that the above is not an inner product.

(3) Let $C[0, 1]$ be the set of all real continuous functions defined over interval $[0, 1]$. Show that the following $\langle, \rangle$ is an inner product,
$$\langle f, g \rangle = \int_0^1 f(x)g(x)dx, \quad \forall f, g \in C[0, 1].$$

(4) Let $F[0, 1]$ be the set of all real bounded integrable functions defined over interval $[0, 1]$. Show that the following $\langle, \rangle$ is not an inner product,
$$\langle f, g \rangle = \int_0^1 f(x)g(x)dx, \quad \forall f, g \in F[0, 1].$$

(5) Let us use the usual measure on interval $[0, 1]$. Let $M[0, 1]$ be the set of all bounded and measurable functions defined over interval $[0, 1]$. Show that the following $\langle, \rangle$ is not an inner product, where the integral sign indicates the Lebesgue integration,
$$\langle f, g \rangle = \int_0^1 f(x)g(x)dx, \quad \forall f, g \in M[0, 1].$$

(6) Let us use the usual measure on interval $[0, 1]$. Let $\overline{M}[0, 1]$ be the quotient set of all bounded and measurable functions defined over interval $[0, 1]$ under the equivalent relation $\approx$ which is defined as $f \approx g$ iff $f - g$ equals 0 outside a measure 0 set. Prove that $\overline{M}[0, 1]$ is a vector space, and $\overline{M}[0, 1]$ is an inner product space with respect to the following $\langle, \rangle$, where the integral sign indicates the Lebesgue integration,
$$\langle f, g \rangle = \int_0^1 f(x)g(x)dx, \quad \forall f, g \in \overline{M}[0, 1].$$

4.2   Gram–Schmidt Theorem. Hooke's Law.
Least Square Approximation

Definition   A set $\{v_i\}_{i \in I}$ is said to be an orthonormal set iff (1) $|v_i| = 1$, $\forall i$. (2) $\langle v_i, v_j \rangle = 0$, $\forall i \neq j$ (or simply $\langle v_i, v_j \rangle = \delta_{ij}$ where δ_{ij} is the Kronecker *delta*). ∎

Proposition 4.2.1:   *Let $u_1, u_2, \ldots, u_n$ be an orthonormal set and U be the subspace spanned by them. Then $u_1, u_2, \ldots, u_n$ forms a basis for U. For any vector $v \in V$, the projection of vector v to the subspace U, $P_U(v) = \langle v, u_1 \rangle u_1 + \langle v, u_2 \rangle u_2 + \cdots + \langle v, u_n \rangle u_n$.*

Proof. It suffices to show that $v - \sum_i \langle v, u_i \rangle u_i$ is $\perp U$, which is obvious. ∎

Furthermore we have the *Gram*[4]*–Schmidt*[5] *process* to construct an orthonormal set,

Theorem 4.2.2:   *Let V be an inner product space and $v_1, v_2, \ldots, v_k, \ldots$ be linearly independent. Then there exists an orthonormal set $u_1, u_2, \ldots, u_k$ such that for any $k = 1, 2, \ldots$, the set*

$$\{u_1, u_2, \ldots, u_k\}$$

is a basis for the subspace spanned by $v_1, v_2, \ldots, v_k$.

Proof. We make an induction for k. For $k = 1$, let $u_1 = v_1/|v_1|$. Let us assume that inductively we have constructed $\{u_1, u_2, \ldots, u_k\}$. If there is no more v_{k+1}, then we are done. Let us assume that there is v_{k+1}. Let

$$w_{k+1} = v_{k+1} - \sum_{i=1}^{k} \langle v_{k+1}, u_i \rangle u_i$$

then w_{k+1} is not 0, otherwise $\{v_1, v_2, \ldots, v_{k+1}\}$ must be a linearly dependent set. We take $u_{k+1} = w_{k+1}/|w_{k+1}|$. ∎

---

[4]Gram, J. Danish mathematician. 1850–1916.
[5]Schmidt, E. Baltic German mathematician. 1876–1959.

Example 4.4 (Least Square Approximation): We shall consider the real field $\mathbb{R}$ only. The least square method was first described by C. Gauss around 1794, his application was about the future location of the newly-discovered asteroid Ceres. Let us consider the simplest case of this method by taking an example from physics. A spring should obey Hooke's[6] law which states that the extension of the spring is proportional to the force, F, applied to it, as

$$p_i = \text{extension} = f(F_i, k) = kF_i$$

where k is the *constant of elasticity* of the material. The above equations may be over-determined and have no common solution for k. We may add unknown ϵ_i to make them consistent and solvable,

$$p_i = kF_i + \epsilon_i, \ \forall i = 1, \ldots, n.$$

We may want to find the solution with minimal square sum of errors ϵ_i. This is the *least square method*, i.e., let $S^2 = \sum_i \epsilon^2$ and

$$\min S^2 = \min \left\{ \sum_i \epsilon_i^2 \right\} = \min \sum (p_i - kF_i)^2.$$

Using *Calculus*, we have

$$\frac{dS^2}{dk} = \sum_i -2F_i(p_i - kF_i) = 2 \left(\sum_i k(F_i^2) - F_i p_i \right) = 0.$$

Therefore the best estimate of k is the value

$$k = \frac{\sum_i F_i p_i}{\sum_i F_i^2}.$$

The above equation can be rewritten as

$$\begin{pmatrix} F_1 \\ F_2 \\ . \\ F_n \end{pmatrix} (k) = \begin{pmatrix} p_1 - \epsilon_1 \\ p_2 - \epsilon_2 \\ . \\ p_n - \epsilon_n \end{pmatrix}.$$

Or we simply write as $Fk = p - \epsilon$, we may even drop ϵ and get an over-determined system of equations $Fk = y$. Our final solution

---

[6]Hooke, R. English natural philosopher, architect and polymath. 1635–1703.

of the over-determined system of equations is $F^T F k = (\sum_i F_i^2)k = F^T p = \sum_i F_i p_i$.

In general, we have a system of equations as follows, $Ax = [c_1, c_2, \ldots, c_n]x = b$. This system of equations has a solution set iff $b \in$ the column subspace spanned by $c_1, c_2, \ldots, c_n$. In general, the above requirement may not be satisfied. Theoretically, then we have no solution. Even in this impossible case, we want to have the *best possible* solution set. For instance, we are measuring the trajectory of a missile which should be $h = a_0 t^3 + a_1 t^2 + a_2 t + a_3$. After 100 times of collecting the height (h) at time $t = t_1, t_2, \ldots, t_{100}$, we get 100 equations in the four unknowns a_0, a_1, a_2, a_3. Due to the measurement errors, we are likely to have an over-determined system of equations. We want the best approximation. We have the following picture, where the proj(b) is the orthogonal projection of b to the column subspace,

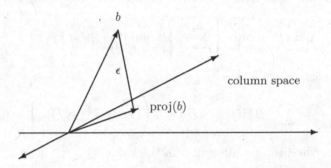

We may easily see that the system of equations $Ax = b$ is over-determined iff $b \notin$ the column space. In this case, we may introduce a vector $\epsilon = [\epsilon_1, \ldots, \epsilon_m]^T$ as $Ax = b + \epsilon$ to make the system of equations solvable. We want to find x with minimal of the sum of the squares $\sum_i \epsilon_i^2$. Recall that for any vectors v, u, $v^T u = u^T v$. Let $x = [x_1, x_2, \ldots, x_n]^T$. Using *Calculus*, we minimize $\epsilon^T \epsilon = \sum_i \epsilon_i^2$ by setting $\frac{\partial \epsilon^T \epsilon}{\partial x_i} = 0$ for $i = 1, 2, \ldots, n$ which gives us for all i the following equation,

$$\frac{\partial([x_1, x_2, \ldots, x_n]A^T - b^T)(A[x_1, x_2, \ldots, x_n]^T - b)}{\partial x_i} = 0.$$

We shall expand the above as

$$([0, 0, \cdot, 1, 0, \cdot, 0]A^T)(Ax - b) - (x^T A^T - b^T)(A[0, 0, \cdot, 1, 0, \cdot, 0]^T) = 0.$$

Let $A = [c_1, c_2, \ldots, c_n]$ where $c_1, c_2, \ldots, c_n$ are column vectors of A. Then the above equation can be rewritten as $(c_i^T)(Ax - b) + (Ax - b)^T c_i = 0$ for $i = 1, 2, \ldots, n$. Since for any two vectors v, u, we always have $v^T u = u^T v$, then we have $(c_i^T)(Ax - b) = (Ax - b)^T c_i$. Therefore, we have $c_i^T(Ax - b) = 0$, for $i = 1, 2, \ldots, n$, and thus $A^T(Ax - b) = A^T \epsilon = 0$. In other words, we have $A^T Ax = A^T b$.

To summarize, we replace an over-determined system of equations $Ax = b$ by the least square approximation

$$A^T Ax = A^T b$$

and the error vector ϵ is characterized by $A^T \epsilon = 0$ or $\epsilon \perp$ column space of A and $b + \epsilon \in$ the column space of A, or $b + \epsilon$ is the orthogonal projection of b to the column space of A. ∎

Example 4.5 (Trigonometric Series): Let us assume that the field is the real field $\mathbb{R}$. We dream of replacing a general *good* function by a sequence of numbers. The theory of trigonometric series is one of them.

The topic of trigonometric series began in 1807 by J. Fourier[7] to treat the 1-dimensional *heat equations*

$$\frac{\partial h(x, t)}{\partial t} = k \frac{\partial^2 h(x, t)}{(\partial x)^2}$$

in the case of the 3-dimensional space, the equation is of the form

$$\frac{\partial h}{\partial t} = k\Delta(h)$$

and k is called the *thermal diffusivity*.

Now the study of heat equation extends from physics to social study. Many actions of the society seems to be random like the actions of atoms under heat. For instance, the transmission of

---

[7]Fourier, J. French mathematician and physicist. 1768–1830.

diseases, passing of information, the motion of stocks, etc. may be all random, hence may be governed by the heat equation.

Let us come back to the solutions of 1-dimensional heat equation. Let us use the method of the separation of variables: let $h(x,t) = f(t)g(x)$, and transform the differential equation to

$$\frac{df(t)}{dt}g(x) = k\frac{d^2g(x)}{(dx)^2}f(t).$$

We further rewrite the equation as two equations,

$$\frac{df(t)}{dt} = -k\lambda f(t)$$

$$\frac{d^2g(x)}{(dx)^2} = -\lambda g(x).$$

Let us consider the problem as over the interval $[-\pi, \pi]$ and under the initial condition,

$$h(x,0) = \alpha(x)$$

and the boundary conditions,

$$h(-\pi, t) = 0 = h(\pi, t) \quad \forall t > 0$$

then one may take $\lambda = n^2$, $g(x) = a_n \sin(nx)$, and $f(t) = e^{-n^2kt}$ as solutions where n may be assumed to be positive integers. We conclude that the solutions are of the form

$$h(x,t) = \sum_{n=1}^{\infty} a_n(\sin(nx))e^{-n^2kt}$$

where

$$a_n = (1/2\pi)\int_{-\pi}^{\pi} \alpha(x)\sin(nx)dx.$$

If we change the initial and boundary conditions, then we have to use $\cos(mx)$. Then we have the following important identities from

Calculus,

$$\int_{-\pi}^{\pi} \cos(mx)\cos(nx)dx = \pi\delta_{mn}, \quad \int_{-\pi}^{\pi} \sin(mx)\sin(nx)dx = \pi\delta_{mn},$$

$$\int_{-\pi}^{\pi} \cos(mx)\sin(nx)dx = 0.$$

Therefore $\{1/(2\pi),\ldots,(1/\sqrt{\pi})\sin(nx),(1/\sqrt{\pi})\cos(nx),\ldots\}$ is a set of orthonormal elements. Furthermore, let us define the space of trigonometric series $T([-\pi,\pi])$ to be

$$T([-\pi,\pi]) = \left\{ c + \sum_n a_n \sin nx + b_n \sum_n \cos nx : \right.$$

$$\left. c^2 + \sum_n a_n^2 + \sum_n b_n^2 < \infty \right\}.$$

In the complex numbers $\mathbb{C}$ case we have the Euler formula, where the pure imaginary number i is defined as $i = \sqrt{-1}$,

$$\begin{bmatrix} e^{inx} = \cos nx + i\sin nx \\ e^{-inx} = \cos nx - i\sin nx \end{bmatrix}.$$

We may replace the pair $\{\sin nx, \cos nx\}$ by the pair $\{e^{inx}, e^{-inx}\}$, and we define the complex trigonometric Fourier series as

$$f = c + \sum_{j=-\infty}^{\infty} \langle f, e^{ijx}\rangle e^{ijx}.$$

In 1966, L. Carleson[8] showed that the trigonometric functions in the real case and the exponential functions in the complex case, represent any $f \in L_2([-\pi,\pi])$ almost everywhere. ∎

Example 4.6 (Discrete Fourier Transform and Fast Fourier Transform): This process is due to J. Cooley[9] and J. Tukey[10] (1965). The Discrete Fourier Transform of $f(x)$ is an expression

---

[8]Carleson, L. Swedish mathematician. 1928–.
[9]Cooley, J. American mathematician. 1926–2016.
[10]Tukey, J. American mathematician. 1915–2000.

$f(x) = \sum_{j=0}^{n-1} a_j e^{ijx}$ where the coefficients a_j are determined by the values of $f(x)$ when $x = 0, 2\pi/n, 2(2\pi/n), \ldots, (n-1)(2\pi/n)$ as follows with $e^{i2\pi/n} = w$,

$$f(1) = v_0 = a_0 + a_1 + \cdots + a_{n-1}$$
$$f(w) = v_1 = a_0 + a_1 w + \cdots + w^{n-1} a_{n-1}$$
$$\vdots$$
$$f(w^{n-1}) = v_{n-1} = a_0 + a_1 w^{n-1} + \cdots + w^{(n-1)^2} a_{n-1}.$$

Note that the coefficients a_j is not the Fourier coefficients in the classical Fourier series. Rather what we have is a relation between coefficients $\{a_j\}$ and values $\{v_j\}$.

If we replace e^{ix} by y (we have $y = 0, w, w^2, \ldots, w^{n-1}$), then $f = a_0 + a_1 y + \cdots + a_{n-1} y^{n-1}$. From this point of view, the *Discrete Fourier Series* is really about the relation between the coefficients and values of a polynomial, in the spirit of Lagrange.[11] The discrete Fourier transform is very important in modern computation mathematics. Maybe the reader will wonder why bother to transfer a set of values to a set of coefficients, then transmit the set of coefficients, and then translate back to a set of values. Is it a complete waste? The major points are (1) there are coding (i.e., self-correcting codes) methods on the coefficients (cf. Section 3, Chapter 1, Reed–Solomon code), (2) there are differences between the values and coefficients of a polynomial in applications. Major blunders in the transmitting of some picture (values) becomes minute alternations to spreading out to the whole picture (coefficients), thus blunders become mistakes only, (3) applications as *digital signal processing* and *latent semantic analysis* (remember some old *internet search engine*?). There, it can be shown that the first 10% of coefficients carry 90% of information. Therefore, we may only treat a small part of the data. Furthermore, there is a fast method to achieve this transformation, which is the so-called *Fast Fourier Transform*, which was considered by someone to be the most important achievement in computational mathematics in the 20th century.

---

[11]Lagrange, J.-L. Italian mathematician. 1736–1813.

We have the following equation,

$$\begin{pmatrix} 1 & 1 & \cdot & 1 \\ 1 & w & \cdot & w^{n-1} \\ \cdot & & & \cdot \\ \cdot & & & \cdot \\ 1 & w^{n-1} & \cdot & w^{(n-1)^2} \end{pmatrix} \begin{pmatrix} a_0 \\ a_1 \\ \cdot \\ \cdot \\ a_{n-1} \end{pmatrix} = \begin{pmatrix} v_0 \\ v_1 \\ \cdot \\ \cdot \\ v_{n-1} \end{pmatrix}.$$

Let us call the $n \times n$ matrix F_n. Note that w is a primitive nth root of unity and $(1/\sqrt{n}F_n)^{-1} = (1/\sqrt{n})\overline{F_n}$. The trick is that if $n = m\ell$, then this process can be sped up. Say $n = 2m$, then as noticed by Cooley and Tukey (1965), we shall rewrite $a' = (a_0, a_2, a_4, \ldots)$ and $a'' = (a_1, a_3, a_5, \ldots)$ and $v' = F_m a'$, $v'' = F_m a''$, we have for $j = 0, 1, \ldots, m - 1$ the following,

$$v_j = (a_0 + w^{2j} a_2 + \cdots) + w^j (a_1 + w^{2j} a_3 + \cdots),$$
$$v_{m+j} = (a_0 + w^{2m+2j} a_2 + w^{4m+4j} a_4 + \cdots) + w^{m+j} (a_1 + \cdots).$$

Note that $(w^2)^m = 1$, $w^m = -1$, we have

$$v_j = v'_j + w^j v''_j, \qquad j = 0, 1, \ldots, m - 1.$$
$$v_{m+j} = v'_j - w^j v''_j, \quad j = 0, 1, \ldots, m - 1.$$

Let $G(m)$ be the number of steps of computations as indexed by m. Then to find $G(n) = G(2m)$, we have to compute v'_j, v''_j and $w^j v''_j$ for $j = 0, \ldots, m - 1$. To compute v'_j and v''_j for $j = 0, \ldots, m - 1$, inductively we need $2G(m)$ steps. Therefore, we have $G(n) \leq 2G(m) + m$. In general, if $n = 2^\ell$ then we may replace F_n by two copies of $F_{2^{\ell-1}}$ and four copies of $F_{2^{\ell-2}}$ and finally n copies of F_1 which is trivial. It is not hard to see the following proposition,

Proposition 4.2.3: *If* $n = 2^\ell$, *then* $G(n) \leq (1/2)n \log_2 n$.

Proof. The proposition is true for $\ell = 0$. We shall use induction. Let $m = 2^{\ell-1}$, $G(2^{\ell-1}) \leq (1/2)2^{\ell-1}(\ell - 1)$. Then we have $G(n) \leq 2G(2^{\ell-1}) + 2^{\ell-1} = (1/2)2^\ell \ell - e^{\ell-1} + e^{\ell-1} = (1/2)2^\ell \ell$. ∎

Ordinarily we need n^2 multiplications, say $n = 2^{12} = 4,096$, then ordinary multiplication requires $(2^{12} - 1)^2$ steps, and the fast Fourier transformation needs $6 \cdot 2^{12}$ the ratio of these two numbers is 682, so the fast Fourier transform is 682 times faster. ∎

Example 4.7: Given the coefficients $a_0, a_1, \ldots, a_{n-1}$. Let us consider the evaluation of $f(x)$ at $x = 1, \omega, \ldots, \omega^{n-1}$ for $n = 2, 4, 8$.

(1) Let $n = 2$. Then we have $f(x) = a_0 + a_1 x$, and we want to evaluate at $x = 1, -1$. Clearly the values are $a_0 + a_1$ and $a_0 - a_1$. So we need no time since only 1 multiplication needs 1 time, not addition nor subtraction.

(2) Let $n = 4$. Then we have $f(x) = a_0 + a_1 x + a_2 x^2 + a_3 x^3$, and we have to evaluate at $x = \pm 1, \pm i$. Let us rewrite $f(x) = (a_0 + a_2 x^2) + x(a_1 + a_3 x^2) = (a_0 + a_2 y) + x(a_1 + a_3 y) = f_0(y) + x f_1(y)$ where $f_i(y)$ are polynomials of degree less than 2 in y to be evaluated at $y = x^2 = \pm 1$. We need no time to find $f_0(y)$ and $f_1(y)$, while we need one multiplication to find $i f_1(-1)$. Therefore we need only one computation to evaluate $f(x)$ at the four points.

(3) Let $n = 8$. Then we have $f(x) = a_0 + a_1 x + a_2 x^2 + a_3 x^3 + a_4 x^4 + a_5 x^5 + a_6 x^6 + a_7 x^7$, and we have to evaluate at $x = \pm 1, \pm \omega, \pm i \pm \omega^3$. Let us rewrite $f(x) = (a_0 + a_2 x^2 + a_4 x^4 + a_6 x^6) + x(a_1 + a_3 x^2 + a_5 x^4 + a_7 x^6) = (a_0 + a_2 y + a_4 y^2 + a_6 y^3) + x(a_1 + a_3 y + a_5 y^2 + a_7 y^3) = f_0(y) + x f_1(y)$ where $f_i(y)$ are polynomials of degree less than 4 in y to be evaluated at $y = x^2 = \pm 1, \pm i$. We need time 2 to find $f_0(y)$ and $f_1(y)$, while we need three multiplications to find $\omega^j f_1(\omega^{2j})$. Therefore we need only five computations to evaluate $f(x)$ at the eight points. ∎

Exercises

(1) Let $P_3(x)$ be all real coefficient polynomials of degree ≤ 3. Start with polynomials $\{1, x, x^2, x^3\}$. Find an orthonormal basis.
(2) Find the projection of x^5 to P_1.
(3) Find the projection of x to the vector space generated by $\{1, (1/\sqrt{\pi}) \sin(x)\}$.
(4) Let w satisfy the equation $w^n = 1$ and $w \neq 1$. Prove that $w^{n-1} + w^{n-2} + \cdots + w + 1 = 0$.
(5) Let F_4 be the *fast Fourier transformation* matrix. Find vector c with $F_4[1, 1, 1, 1]^T = c$.
(6) Let F_4 be the *fast Fourier transformation* matrix. Find vector c with $F_4 c = [1, 1, 1, 1]^T$.

4.3   Hilbert Space

There are problems of using *Calculus* to an inner product space. The main problem is that a well-behaved sequence may not have a limit. We should reflect on the basic requirement of *analysis*. When we extend the field from the rational numbers $\mathbb{Q}$ to the real numbers $\mathbb{R}$, we need either the *Dedekind*[12] *cuts* or *Cauchy*[13] *sequences*. The set of rational numbers on the real line is full of gaps which makes analytical study of $\mathbb{Q}$ difficult. If we include a real number at the limit for every Cauchy sequence, then we get the real line, which is *complete*. In fact both methods were known in ancient Greece, and we shall treat them as common knowledge. In this book, we will use the method of Cauchy sequences. We recall the following definition,

Definition   Let V be a metric space with metric d, and $S = \{v_1, \ldots, v_i, \ldots\}$ be a sequence. The sequence S is said to be a *Cauchy sequence* if for any $\epsilon > 0$, there is a number M, such that for all $i, j > M$, we have $d(v_i, v_j) < \epsilon$ always.   ∎

Definition   A metric space V is *complete* if every Cauchy sequence $S = \{v_1, \ldots, v_i, \ldots\}$ has a limit, i.e., there is a v such that $d(v_i, v) \mapsto 0$.   ∎

A Hungarian-born American mathematician, von Neumann,[14] gave the following definition at the turn of the 20th century,

Definition   A complete inner product vector space is called a *Hilbert*[15] *space*.   ∎

This definition was the right one, and had been accepted by the community of scholars.

---

[12]Dedekind, J. German mathematician. 1831–1916.
[13]Cauchy, A. French mathematician, engineer and physicist. 1789–1857.
[14]von Neumann, J. Hungarian–American mathematician, physicist, computer scientist and polymath. 1903–1957.
[15]Hilbert, D. German mathematician. 1862–1943.

We have the following definition,

Definition

(1) Let the field K be the real numbers $\mathbb{R}$ or the complex numbers $\mathbb{C}$. Let ℓ_2 be the set of all square summable sequence, i.e.,

$$\ell_2 = \left\{ (s_1, s_2, \ldots, s_n, \ldots) : s_i \in K, \sum_i |s_i|^2 < \infty \right\}.$$

(2) Let $f : [-\pi, \pi] \mapsto \mathbb{R}$ be measurable. We say that f is square integrable if $|f|^2$ is integrable, i.e., if

$$\int_{-\pi}^{\pi} |f|^2 d\mu < \infty.$$

We say two functions f, g are equivalent iff $\{x : (f - g)(x) \neq 0\}$ is a measure zero set. We write $L^2([-\pi, \pi])$ as all equivalent classes of square integrable functions on $[-\pi, \pi]$. ∎

Then we have the following proposition,

Proposition 4.3.1: *The space ℓ_2 of square summable sequences is a vector space. Let $s = (s_1, s_2, \ldots)$, $t = (t_1, t_2, \ldots) \in \ell_2$, and define $\langle s, t \rangle = \sum_i s_i \bar{t}_i$. Then $\langle , \rangle$ is an inner product for ℓ_2, and ℓ_2 is a Hilbert space under this inner product.*

Proof. We first show that ℓ_2 is a vector space. Consider $as = (as_1, as_2, \ldots)$, then we have

$$\sum_i |as_i|^2 = a^2 \sum_i |s_i|^2 < \infty$$

and $s + t = (s_1 + t_1, s_2 + t_2, \ldots)$, we have

$$\sum_i |s_i + t_i|^2 \leq \sum_i (2|s_i|^2 + 2|t_i|^2) \leq \infty.$$

Henceforth ℓ_2 is a vector space. Next we want to show that $\langle s, t \rangle$ is a finite number, therefore it is well-defined. Note that

$$\sum_i |s_i \bar{t}_i| \leq \sum_i (1/2)|s_i|^2 + (1/2)|t_i|^2 < \infty.$$

We conclude that it is well-defined. Now it is easy to check the function $\langle,\rangle$ satisfies the three axioms of the inner product.

Finally we want to show that ℓ_2 is complete under the inner product $\langle,\rangle$.

Let $(s^{(1)}, s^{(2)}, \ldots, s^{(n)}, \ldots)$ be a Cauchy sequence in ℓ_2, where

$$s^{(1)} = (s_1^{(1)}, s_2^{(1)}, \ldots, s_n^{(1)}, \ldots)$$
$$s^{(2)} = (s_1^{(2)}, s_2^{(2)}, \ldots, s_n^{(2)}, \ldots)$$
$$\cdots\cdots\cdots\cdots\cdots$$
$$s^{(n)} = (s_1^{(n)}, s_2^{(n)}, \ldots, s_n^{(n)}, \ldots)$$
$$\cdots\cdots\cdots\cdots\cdots$$

For any index i, we consider the sequence $(s_i^{(1)}, s_i^{(2)}, \ldots, s_i^{(n)}, \ldots)$, we clearly have

$$d(s_i^{(n)} - s_i^{(m)}) \le d(s^{(n)} - s^{(m)})$$

then the sequence $(s_i^{(1)}, s_i^{(2)}, \ldots, s_i^{(n)}, \ldots)$ is Cauchy. Since the real number field $\mathbb{R}$ is complete, it has a limit s_i. Therefore we have a sequence $s = (s_1, s_2, \ldots, s_n, \ldots)$. We wish to show that

$$s = \lim_{n \mapsto \infty} s^{(n)} \in \ell_2.$$

Let us consider the following inequalities,

$$d(s^{(n)}, s^{(m)}) \le \epsilon \quad \text{for all } n, m \ge N.$$

We may take $n = N$, and have for any integer M the following

$$\sum_{i=1}^{M} |s_i^{(N)} - s_i^{(m)}|^2 \le \epsilon^2 \quad \text{for all } m \ge N.$$

We fix m and let N go to ∞, then we have

$$\sum_{i=1}^{M} |s_i - s_i^{(m)}|^2 \le \epsilon^2.$$

Since M is arbitrarily selected, we have

$$d(s, s^{(m)}) \le \epsilon.$$

It is easy to check that

$$|s| = |s_N + s - s_N| \le |s - s_N| + |s_N| \le |s_N| + \epsilon.$$

Therefore the norm of s is bounded and $s \in \ell_2$ and the limit of the sequence is s. ∎

Remark: It is known that $L^2([-\pi, \pi])$ is a Hilbert space. For further discussion, see the next section. ∎

Later on we will show that $\mathbb{R}^n$'s, $\mathbb{C}^n$'s are Hilbert spaces. Let us consider $\mathbb{C}^\infty = \oplus_{i \in N} \mathbb{C}$ where N is the set of natural numbers. Note that $\mathbb{C}^\infty = \oplus_{i \in N} \mathbb{C} = \{(a_1, a_2, \ldots, 0, 0, \ldots) : a_i \in \mathbb{C}\}$. It is a vector space. Let us define the inner product in $\mathbb{C}^\infty = \oplus_{i \in N} \mathbb{C}$ for two vectors $v = (a_1, a_2, \ldots, 0, 0, \ldots)$, $u = (b_1, b_2, \ldots, 0, 0, \ldots)$ as

$$\langle v, u \rangle = \sum_i a_i \bar{b}_i.$$

Then $\mathbb{C}^\infty$ is an inner product space and a subspace of ℓ^2. Note that the sequence $\{s_n\} = \{(\overbrace{1, 1/2^1, \ldots, 1/2^{n-1}}^{n}, 0, 0, \ldots)\}$ clearly goes to $(1, 1/2, \ldots, 1/2^n, \ldots)$ which is not in $\mathbb{C}^\infty$. Hence $\mathbb{C}^\infty$ is not complete and not a Hilbert space.

Exercises

(1) Let $\mathbb{R}^n$ be an n-dimensional vector space equipped with the usual inner product. Show that $\mathbb{R}^n$ is a Hilbert space.

(2) Let $\mathbb{C}^n$ be an n-dimensional vector space equipped with the usual inner product. Show that $\mathbb{C}^n$ is a Hilbert space.

(3) Find a function defined over $[0, 1]$ which is integrable but not square integrable.

(4) Let us consider $\mathbb{C}^\infty \subset \ell^2$. Show that $(1, 1/2!, 1/3!, \ldots, 1/n!, \ldots) \in \ell^2$, and there is no closest element in $\mathbb{C}^\infty$ to $(1, 1/2!, 1/3!, \ldots, 1/n!, \ldots)$.

(5) Let us consider $\mathbb{C}^\infty \subset \ell^2$. Show that $\{y : y \in \ell^2, \langle x, y \rangle = 0 \ \forall x \in \mathbb{C}^\infty\}$ is the set $\{0\}$.

4.4 Orthogonal Complement

We have the following definition,

Definition We write $v^\perp = \{u : \langle v, u \rangle = 0\}$ for any vector v and $U^\perp = \{v : \langle v, u \rangle = 0, \forall u \in U\}$ for any subspace U. Both $v^\perp, U^\perp$ are subspaces. ∎

The following theorem is interesting, we will use some elementary concepts of *topology* and prove them in this *book*,

Theorem 4.4.1: *In a complete metric space V, a subset S is complete iff it is closed.*

Proof. ($\Longleftarrow$) If S is closed, then the complement of S, $V \backslash S$ is open, i.e., for any point $p \in V \backslash S$, there is an open ball of radius ϵ, $B(p, \epsilon)$ with the intersection of it with S the empty set. Therefore any Cauchy sequence of elements in S will be disjoint with $B(p, \epsilon)$ eventually. Hence p cannot be the limit point. Since the limit point exists, it must be in S.

($\Longrightarrow$) Let p be a point in the complement of S. If for every small ϵ, there is a point $q_n \in S \cap B(p, \epsilon)$, then there is a sequence of elements $q_n \in S$ which goes to a point p outside S, contradicting to our assumption that S is complete. Therefore there must be an open ball $B(p, \epsilon)$ which is outside S. Hence S is closed. ∎

Note that a set S is said to be *convex* if $v, u \in S$, $0 \le \alpha \le 1$, then $\alpha v + (1 - \alpha)u \in S$.

Proposition 4.4.2: *Any subspace U of an inner product space is convex.*

Proof. Clearly any subspace, finite or infinite dimensional, is convex. ∎

Lemma 4.4.3: *Given a non-empty closed subspace U of a complete inner product space V. Let $v \in V$, $d = minimal\{d(v, u) : u \in U\}$ and $u_1, u_2, \ldots$ be a sequence in U. If $d(v, u_i)$ is a Cauchy sequence, and goes to d, then $u_1, u_2, \ldots$ is a Cauchy sequence with limit $u \in U$ such that $d(v, u) = d$.*

Proof. If $u_n = u_{n+1} = \ldots$, then we just take $u = u_n$. Otherwise we may assume that since the sequence $d(v, u_i)$ goes down to d which is minimal possible, i.e., given $\epsilon > 0$, there is a k for any $n \geq k$, we have $d + \epsilon > d(v, u_n) \geq d$. Then for any $n, m > k$, we have a triangle formed by the tips of the three vectors v, u_n, u_m, it is not hard to see that the length ℓ of the segment from the tip of vector v perpendicular to the line connecting the tips of u_m, u_n as in the following diagram,

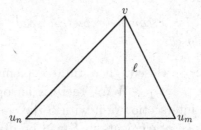

Since the set is convex, the projection of the tip to the line segment connecting the tips of u_n, u_m is in S. We have

$$\left[\begin{aligned} d^2 \leq \ell^2 &\leq \max\{d(v, u_n)^2, d(v, u_m)^2\} - ((1/2)d(u_n, u_m))^2 \\ &\leq (d + \epsilon)^2 - ((1/2)d(u_n, u_m))^2 \end{aligned} \right].$$

We have,

$$d(u_n, u_m)^2 \leq 4(2d\epsilon + \epsilon^2).$$

Therefore $\{u_i\}$ is a Cauchy sequence. According to our assumption, there is a limit u. It is easy to see that $d(v, u) = d$. ∎

We have the following proposition,

Proposition 4.4.4: *Let V be complete and S be a non-empty closed and convex set, and $v \in V$ a vector, then there is a unique vector $u \in S$, such that $|v - u|$ is the minimal of $\{|v - u| : u \in S\}$.*

Proof. The existence of u follows from the preceding lemma. We shall concentrate to prove the uniqueness. If there are two vectors with property, let them be u, u'. We imitate the proof of the preceding lemma, it is easy to see some in-between vector $u'' = \alpha u + (1 - \alpha)u'$ will have a smaller distance to v. A contradiction. ∎

Example 4.8: Not every subspace is closed. Note that it follows from the discussion of Section 4.3 that $\oplus_{i=1}^{\infty}\mathbb{C}$ is an incomplete subspace of ℓ_2, and not closed. ∎

Let U be a closed subspace (for instance, U is a finite dimensional subspace) of V, then we have

Proposition 4.4.5: *Assume that V is complete (i.e., V is a Hilbert space). The closest vector u between v and U which is a closed subspace of V is characterized by $u \in U$, $(v - u) \perp U$.*

Proof. ($\Longrightarrow$) We have a unique $u \in U$, if $v - u \not\perp U$, then there is a unit vector $w \in U$, such that $a = \langle w, v - u \rangle \neq 0$, then we have $\langle w, v - u - aw \rangle = a - a = 0$ and $|v - u|^2 = \langle v - u - aw + aw, v - u - aw + aw \rangle = |v - u - aw|^2 + |aw|^2$. Therefore $|v - u - aw| < |v - u|$ contradict to the minimum of $|v - u|$.

($\Longleftarrow$) Let $w \in U$ be any non-zero vector. We have $v - w = v - u + u - w$. By the Pythagoras theorem, we have $|v - w|^2 = |v - u|^2 + |u - w|^2$. It is not hard to show $|v - w| \geq |v - u|$. ∎

One of the possible applications of the preceding proposition is the best approximate in a closed subspace of a vector. We have the following example,

Example 4.9: Let us consider the Hilbert space $L^2([0, 1])$ which is the equivalent classes of all square integrable functions over $[0, 1]$. We want to find the closest function in the subspace U generated by $1, x$ of the function x^5. Note that U is finite dimensional, hence a Hilbert subspace. According to the preceding proposition, the closest function is the projection of x^5 to U. We have the following computations,

- $\langle 1|1 \rangle = \int_0^1 1^2 dx = 1$.
- $\langle x|1 \rangle = \int_0^1 x \cdot 1 dx = 1/2$.
- $\langle x - 1/2|1 \rangle = \int_0^1 x - 1/2 dx = 0$.
- $\langle (x - 1/2)|(x - 1/2) \rangle = 1/12$.

Therefore the projection of x^5 to U is $\langle x^5|1\rangle 1 + \langle x^5|\sqrt{12}(x - 1/2)\rangle\sqrt{12}(x-1/2) = (5/7)x-4/21$ which is the closest approximation of x^5. ∎

Proposition 4.4.6: *Assume that V is complete. If U is a subspace of V, then $U^\perp$ is a closed subspace of V. Furthermore, if U is closed, then $(U^\perp)^\perp = U$ and $V = U \oplus U^\perp$. If U is closed, $U^\perp$ is called the orthogonal complement of U.*

Proof. Let $w_i, w_2, \ldots$ be a Cauchy sequence in $U^\perp$ with limit w. We have $|w - w_i| \mapsto 0$. By the Cauchy–Bunyakovsky–Schwarz inequality, we have $|\langle w - w_i, u\rangle| \leq |w - w_i||u|$ for any $u \in U$. Therefore $|\langle w, u\rangle - \langle w_i, u\rangle| = |\langle w - w_i, u\rangle| \mapsto 0$ for any $u \in U$. Hence $\langle w, u\rangle = 0$ and $w \in U^\perp$. In other words, $U^\perp$ is complete and closed.

If U is closed, by the preceding proposition, any vector $v \in V$ can be written as $v = u + (v - u)$, where $u \in U$, $v - u \in U^\perp$. It follows that $V = U \oplus U^\perp$ and $(U^\perp)^\perp = U$. ∎

Definition If U is a closed subspace of a Hilbert space V, every vector $v \in V$ can be written as $v = u + w$ with $u \in U$, $w \in U^\perp$. The map $\pi(v) = u$ is called the *projection*, P_U, of V to U. ∎

Proposition 4.4.7: *We always have $P_U^2 = P_U$.*

Proof. It is obvious. ∎

Proposition 4.4.8: *Let U, W be two closed subspaces. Then $P_U P_W = 0 \iff U \perp W$. Furthermore in this case, $P_W P_U = 0$ and $P_U + P_W = P_{U+W}$.*

Proof. ($\implies$) We know image(P_W) $= W$. The equation $P_U P_W = 0$ means $P_U(W) = 0$ which implies $W \subset U^\perp$ which is to say $U \perp W$.

($\impliedby$) We know that $P_W(v) \in W$ for any $v \in V$. Therefore $P_W(v) \in U^\perp$ and $P_U P_W(v) = 0$ for any $v \in V$ which means $P_U P_W = 0$.

Furthermore, $U \perp W \iff W \perp U$. Therefore we have $P_W P_U = 0$. Let us consider $(P_U + P_W)^2 = P_U + P_W$. The range of $P_U + P_W$ is clearly $U + W$. It is not hard to see that $P_U + P_W = P_{U+W}$. ∎

Exercises

(1) Show that in $\mathbb{R}^n$, a vector $u = (b_1, b_2, \ldots, b_n)^T$ has length $\sqrt{b_1^2 + b_2^2 + \cdots + b_n^2}$.

(2) Let V, U be two inner product spaces. Show that $V \otimes U$ is an inner product space by defining

$$\langle v_1 \otimes u_1, v_2 \otimes u_2 \rangle = \langle v_1, v_2 \rangle \langle u_1, u_2 \rangle$$

and then generalize to others by linearity.

(3) Show the following *parallelogram identity*,

$$|v + u|^2 + |v - u|^2 = 2(|v|^2 + |u|^2).$$

(4) Prove Proposition 4.4.7.

(5) Let U, W be two closed subspaces of a Hilbert space V. Show that $P_U P_W$ is a projection, iff P_U and P_W commute, and in this case $P_U P_W = P_{U \cap W}$.

(6) If U is a closed subspace, then $(U^\perp)^\perp = U$.

(7) Give an example to show that $(U^\perp)^\perp \neq U$.

(8) Let V be a complete inner product space. Show that $V \neq U \oplus U^\perp$ if U is not closed.

(9) Find an example of a non-closed subspace U of a complete inner product space V, and a vector $v \in V$ such that there is no vector $u \in U$ with $|v - u| = minimal\{|v - u_i| : u_i \in U\}$.

(10) Let V be inner product space, and U an invariant subspace under a linear operator T. Show that the closure $\hat{U}$ of U is an invariant subspace under T.

(11) If S, T are non-zero linear operators on V, and $ST = 0$, then the nullspace of S is non-trivial invariant subspace under both S, T.

(12) Let the set of all formal power series $R[[x]] = \{\sum_{i=m}^{\infty} a_i x^i : m \geq 0, a_i \in R\}$ be a topological space. We define the function $d(f, g) = 2^{-\text{ord}(f-g)}$. Show that this function $d(f, g)$ is a distance. Under this definition of distance, show that every triangle is an isosceles, every interior point of a circle is a center.

(13) Let V be a complex vector space. A map $\phi : V \times V \mapsto C$ is said to be sesquilinear if

$$\phi(v + u, w + z) = \phi(v, w) + \phi(v, z) + \phi(u, w) + \phi(u, z),$$

$$\phi(ay, bu) = \bar{a}\bar{b}\phi(v, u).$$

Show that the inner product $\langle, \rangle$ is sesquilinear.

(14) Furthermore a sesquilinear form ϕ is said to be hermitian iff

$$\phi(v, u) = \overline{\phi(u, v)}.$$

Show that the inner product is hermitian, and not every hermitian form is an inner product.

(15) If V is finite dimensional, then relative to any basis $\{e_i\}$, the matrix form H of a hermitian form, i.e.,

$$h(v, u) = v^T H \bar{u}$$

is hermitian, i.e., $H = \overline{H^T}$. Show that the inner product is hermitian.

(16) Let T be a linear operator on an inner product space V. Show that the nullspace of T is an invariant subspace under any linear operator S which commutes with T.

(17) Let T be a linear operator on an inner product space V. Show that the closure of the range of T, $\widehat{R(T)}$, is an invariant subspace under any linear operator S which commutes with T.

4.5 Adjoint and Self-Adjoint Operator. Uncertainty Principle

Before we introduce *linear operators*, we shall remark on the domain of a linear operator. Many times we have to consider functions, transformations etc., defining on a dense subset of the whole space. For instance, all functions defined over a projective algebraic variety are constants only, therefore the sets of functions defined over the varieties do not tell different projective algebraic varieties apart. We have to consider those functions with the domain of definition dense subsets, then the set of functions will reflect the properties of the projective algebraic varieties. Similarly in the case of inner product spaces, we shall consider linear operators with dense domain of definition. The other reason is in *Quantum Mechanics*, there are many operators defined only on a dense subset. For instance, the *momentum operator* P is as follows,

$$\frac{\hbar}{i}\frac{\partial}{\partial x}$$

where $\hbar$ is the reduced Planck[16] constant and $i = \sqrt{-1}$. It is clear that the momentum operator is not defined over the whole Hilbert space V. Another one is the position operator x.

Example 4.10 ($L^2([M])$): Let us consider another inner product space: the space of all *continuous* complex-valued functions f, g defined over the interval $[0, 1]$ with the inner product defined as

$$\langle f, g \rangle = \int_0^1 f\bar{g}dx.$$

The length is defined as

$$|f|^2 = \int_0^1 |f|^2 dx.$$

It is easy to see that this is an inner product space, while it is not a Hilbert space, because the limit of a Cauchy sequence of continuous functions may not be continuous.

We have to reinterpret all concepts, and so we have to use *Lebesgue*[17] *integral* instead of Riemannian integral. Then there is a problem of the third axiom of inner product; in the case of Lebesgue integral, a function which is non-zero only on a measure zero set will have Lebesgue integral zero, thus a non-zero function may have a zero integration, therefore we have to further define *equivalence* of functions to remedy the situation. The resulting object is named $L^2([M])$ which consists of equivalent measurable functions on the measurable set M for which the Lebesgue integral of the second

---

[16]Planck, M. German physicist. 1918 Nobel Prize in Physics. 1858–1947.
[17]Lebesgue, H. French mathematician. 1875–1941.

power of the absolute value of the function is finite over the set M i.e., the following 2-norm

$$|f|_2 = \left(\int_0^1 |f|^2 d\mu \right)^{1/2} < \infty.$$

The completeness of this space is the so-called *Riesz*[18]–*Fischer*[19] theorem.

Especially interesting is the fact that $L^2(\mathbb{R}^3)$ is used in *Quantum Mechanics* as follows,

(1) We define the *quantum state* as $\psi \in L^2(\mathbb{R}^3)$ with $|\psi| = 1$.
(2) We define the *quantum observables* as the self-adjoint operators A.
(3) The expected value a of an observable A at the state ψ is

$$a = \langle A\psi, \psi \rangle.$$

(4) We have the *standard deviation* from statistics $(\Delta_\psi A)^2 = \langle (A - a)^2 \psi, \psi \rangle$, where a is the expected value of ψ.

We notice that if $A\psi = \lambda\psi$, then λ is real and $\langle A\psi, \psi \rangle = \lambda$. However for two observables A, B, for a given ψ we have,

$$
\begin{aligned}
|\langle (A - a)(B - b)\psi, \psi \rangle| &= |\langle (B - b)\psi, (A - a)\psi \rangle| \\
&\leq |(A - a)\psi||(B - b)\psi| \\
&= |\langle (A - a)^2 \psi, \psi \rangle^{1/2} ||\langle (B - b)^2 \psi, \psi \rangle^{1/2}| \\
&= (\Delta_\psi(A))(\Delta_\psi(B)).
\end{aligned}
$$

Furthermore we have

$$
\begin{aligned}
|\langle (A - a)&(B - b)\psi, \psi \rangle| \\
&\geq |\mathrm{im}\langle \overline{((A - a)(B - b)}\psi - (B - b)(A - a)\psi, \psi) \rangle| \\
&= (1/2)|\langle ((A - a)(B - b)\psi - (B - b)(A - a)\psi, \psi) \rangle| \\
&= (1/2)|\langle (AB - BA)\psi, \psi \rangle|.
\end{aligned}
$$

---

[18] Riesz, F. Hungarian mathematician. 1880–1956.
[19] Fischer, E. Austrian mathematician. 1875–1954.

Therefore we conclude

$$(1/2)|\langle (AB - BA)\psi, \psi \rangle| \leq (\Delta_\psi(A))(\Delta_\psi(B)).$$

We may consider A = the momentum operator and B = the position operator. Then it is easy to see that

$$[AB - BA]\psi = \frac{\hbar}{i}\frac{\partial x\psi}{\partial x} - \frac{x\hbar}{i}\frac{\partial \psi}{\partial x} = \frac{\hbar}{i}\psi$$

$$(1/2)|\langle (AB - BA)\psi, \psi \rangle| = \frac{\hbar}{2} \leq (\Delta_\psi(A))(\Delta_\psi(B)).$$

This shows that if one of $(\Delta_\psi(A))$, $(\Delta_\psi(B))$ gets smaller, then the other one must be bigger. This is the famous *uncertainty principle* of Heisenberg.[20]

In the Rutherford model of the atom, it is pictured as a tiny solar system. However, the motion of the electron in the electromagnetic field of the nucleus will produce electric waves, the electron will lose energy and eventually fall down into the nucleus. This is quite impossible.

The remedy is the model of Heisenberg using the uncertainty principle. We have a static model where the electron is a probability cloud covering the nucleus. The radius of the atom depends on the uncertainty principle. ∎

We have the following definition,

Definition A *linear operator* A of Hilbert space V is a linear map from the domain of A, in symbol $\text{Dom}(A)$, to V, where $\text{Dom}(A)$ is a dense subspace of V. ∎

In this section, we will separate two cases: V is finite dimensional or V is infinite dimensional. If V is finite dimensional, then it is clear that the only dense subspace is the whole space, therefore $\text{Dom}(A) = V$ always. If V is infinite dimensional, there are many beautiful theorems. We will only slightly touch on them; interested readers should study the topic of *Hilbert space* in analysis.

---

[20]Heisenberg, W. German physicist. 1932 Nobel Prize in Physics. 1901–1976.

We shall show that a linear operator A whose domain, $\text{Dom}(A) = V$, is determined by its value for an inner product as follows,

Proposition 4.5.1: *Let V be an inner product space and A a linear operator with $\text{Dom}(A) = V$. Then A is determined by $\langle Av, u \rangle$ for all v, u, i.e., if B is another linear operator with $\text{Dom}(B) = V$, and $\langle Av, u \rangle = \langle Bv, u \rangle$ for all v, u, then $A = B$.*

Proof. Let $C = A - B$. It suffices to show that $C = 0$. Note that $\langle Cv, u \rangle = 0$ for all v, u. Especially it is true for $u = Cv$, and we have $|Cv| = 0 \; \forall v$. Therefore $C = 0$. ∎

Corollary 4.5.2: *We have $A = 0$ iff $\langle Av, u \rangle = 0$ for all $v, u \in V$.* ∎

Let us study the inner product $\langle v, u \rangle$ further. It is linear in the first variable v. Let us rewrite $f_u(v) = \langle v, u \rangle$. Then $f_u \in V^* = \text{Hom}_C(V, C)$, where f_u is a linear functional. Let A be any linear operator, then we define an adjoint A^* of A as follows,.

Definition Given a linear operator A, if there is a linear operator A^* such that $\langle Av, u \rangle = \langle v, A^*u \rangle$, $\forall v \in \text{Dom}(A)$, and $u \in \text{Dom}(A^*)$, then we say that A has adjoint A^*. If $\text{Dom}(A) = \text{Dom}(A^*)$, $A = A^*$, then we say A is *self-adjoint*. ∎

Example 4.11: In the finite dimensional case, we may select an orthonormal basis $\{e_1, \ldots, e_n\}$. Then a linear operator may be identified with its representation as a matrix A. Then the adjoint A^* of A is the transpose of conjugate $\overline{A}^T$. Therefore the adjoint always exists.

(1) Given $\ell^2 = \{(a_1, a_2, \ldots) : \sum_i |a_i|^2 < \infty\}$. We use the usual inner product $\langle v, u \rangle = \sum_i a_i \overline{b_i}$ where $v = [a_1, a_2, \ldots]^T$, $u = [b_1, b_2, \ldots]^T$. Let A be the shifting indices by 1 i.e., $A(a_1, a_2, a_3, \ldots) = (0, a_1, a_2, \ldots)$. Then the adjoint A^* of A is given by $A^*(b_1, b_2, b_3, \ldots) = (b_2, b_3, \ldots)$.

(2) However there may not be an adjoint in the infinite dimensional cases. Let $S = \oplus_i R \subset \ell^2$, where S is the set of all finite long

sequences $(a_1, \ldots, a_n, 0, \ldots)$. We use the inner product in ℓ^2. Fix $z \in S$ non-zero element, and $\xi \in \ell^2 \backslash S$. Define $A(x) = \langle x, \xi \rangle z$. Then we have

$$\langle Ax, y \rangle = \langle \langle x, \xi \rangle z, y \rangle = \langle x, \xi \rangle \langle z, y \rangle = \langle x, \overline{\langle z, y \rangle} \xi \rangle$$

$$= \langle x, A^* y \rangle, \quad A^* y = \overline{\langle z, y \rangle} \xi,$$

$$A^* z = \overline{\langle z, z \rangle} \xi = |z|^2 \xi.$$

Therefore $A^* z$ is not in S. Hence A^* does not exist as an operator of S. ∎

Definition Let V be a Hilbert space, A is linear operator. If $\langle Av, u \rangle = \langle v, Au \rangle$ for all v, $u \in \text{Dom}(A)$, then the operator A is said to be hermitian.[21] If $\langle Av, u \rangle = \langle v, -Au \rangle$ for all v, $u \in \text{Dom}(A)$, then the operator A is said to be skew-hermitian. ∎

There is a fine difference between the self-adjoint operator and hermitian operator concerning the domain of definitions. In the finite dimensional cases, the domain of definition is always the whole space, and self-adjoint = hermitian. We have the following example to illustrate the difference in the domains of definition,

Example 4.12: There is a difference between the self-adjoint operator and hermitian operator in the infinite dimensional cases. Let us consider $L^2([a, b])$ over the complex number field $\mathbb{C}$. Let us consider the following operator

$$A(\phi) = \frac{1}{i} \frac{\partial \phi}{\partial x}$$

with the boundary condition $\phi(a) = \phi(b) = 0$. We have

$$\langle A\phi, \psi \rangle = \int_a^b A(\phi) \overline{\psi} dx$$

$$= \phi \overline{\psi} \big|_a^b - \frac{1}{i} \int_a^b \phi \frac{\partial \overline{\psi}}{\partial x}$$

$$= \langle \phi, B\psi \rangle$$

---

[21] Hermite, C. French mathematician. 1822–1901.

where B is a linear operator as follows,

$$B(\psi) = \frac{1}{i}\frac{\partial\psi}{\partial x}$$

and without the boundary condition. Therefore we may replace B by A, and verify $\langle Av, u \rangle = \langle v, Au \rangle$, i.e., A is hermitian, while $\mathrm{Dom}(A)$ is a proper subset of $\mathrm{Dom}(B)$, and A is not self-adjoint.   ∎

Remark: In the finite dimensional complex case, a matrix A is *self-adjoint* iff it is *hermitian*. In the finite dimensional real case, a matrix A is *self-adjoint* iff it is *hermitian* and iff it is symmetric.   ∎

For the beauty of nature, the important real matrices are symmetric. It is the high point of classical linear algebra to show that *"every symmetric real square matrix is diagonalizable"*. In the Dirac[22]– von Neumann[23] formulation of *Quantum Mechanics*, all physical observables such as position, momentum, angular momentum and spin are all represented by self-adjoint operators on a Hilbert space. Therefore it is significant to study self-adjoint linear operators. Moreover, the self-adjoint operator is an important topic in classical analysis in itself. We shall take a deeper look at the self-adjoint operator.

Given any A, our previous example shows that there may not be any A^*, while we will show that in some cases there is at most one adjoint. We have,

Proposition 4.5.3: *For any linear operator A with $\mathrm{Dom}(A) = V$, if A^* exists for $\mathrm{Dom}(A^*) = V$, then there is at most one adjoint A^*.*

Proof. Let B, C be both adjoint of A. Then we have $\langle Av, u \rangle = \langle v, Bu \rangle = \langle v, Cu \rangle$. Or $\langle v, Bu - Cu \rangle = 0$ for all v, u. Let $v = Bu - Cu$. Then we get $|Bu - Cu| = 0$, for all u. Which means $B = C$.   ∎

---

[22]Dirac, P. A. English physicist. 1902–1984.
[23]von Neumann, J. Hungarian–American mathematician. 1903–1957.

In many applications, they are concentrated on the eigenvalues for a self-adjoint operator. Let us define,

Definition The set σ_p of the *point spectrum* of A is consisted of λ such that $\lambda I - A$ is not an injective map on V. ∎

Remark: We have a complete definition of the spectrum $\sigma(A) = \sigma_p \cup \sigma_c \cup \sigma_r$, where σ_p is defined in the previous definition, while

$$\sigma_r = \text{residual spectrum} = \{\lambda : \lambda I - A \text{ injective, does not have a dense image}\}$$

$$\sigma_c = \text{continuous spectrum} = \{\lambda : \lambda I - A \text{ injective, with dense image, while non-invertible}\}.$$ ∎

Remark: In the finite dimensional case, an injective linear operator must be surjective and thus invertible. Therefore $\sigma_c = \sigma_r = \emptyset$. Let V be general. Let $\lambda \in \sigma_p$. Then there exists $x \neq 0$ such that

$$(\lambda I - A)x = 0 \iff Ax = \lambda x$$

which is the old definition of eigenvalue and eigenvector (cf. Definition of Section 2.10). We shall name any element $\lambda \in \sigma_p$ as an eigenvalue of A. ∎

Theorem 4.5.4: *Let λ be an eigenvalue of a self-adjoint operator, then λ must be real.*

Proof. Let v be an associated eigenvector of λ, i.e., $Av = \lambda v$, $v \neq 0$. Then we have $\lambda|v|^2 = \lambda\langle v, v \rangle = \langle \lambda v, v \rangle = \langle Av, v \rangle = \langle v, Av \rangle = \overline{\langle Av, v \rangle} = \overline{\lambda}\langle v, v \rangle = \overline{\lambda}|v|^2$, and we conclude that $\lambda = \overline{\lambda}$. Therefore λ must be a real number. ∎

The above theorem depends on the existence of the eigenvalue λ. The following example shows that there may not be any eigenvalue for a self-adjoint operator,

Example 4.13: Let us define an inner product in the real polynomial ring $\mathbb{R}[x]$ as

$$\langle f, g \rangle = \int_{x=0}^{x=1} f(x)g(x)dx.$$

Let

$$Af(x) = xf(x).$$

Then it is easy to see that $A^* = A$ and A has no eigenvalue. ∎

We have the following theorem for the general cases.

Proposition 4.5.5: *Let A be a self-adjoint operator, and $\lambda \neq \mu$ two distinct eigenvalues (they must be both real) of A, moreover let v (respectively, u) be an eigenvector of A associated with λ (respectively, with μ), then v, u are perpendicular to each other, i.e., $\langle v, u \rangle = 0$.*

Proof. We have $\lambda \langle v, u \rangle = \langle Av, u \rangle = \langle v, Au \rangle = \langle v, \mu u \rangle = \overline{\mu} \langle v, u \rangle = \mu \langle v, u \rangle$. Therefore $\langle v, u \rangle = 0$. ∎

Definition A subspace U is invariant under A if $A(U) \subset U$. ∎

Proposition 4.5.6: *Let A be a self-adjoint operator on an inner product space V with $\mathrm{Dom}(A) = V$. If U is an invariant subspace of A, then the restriction of A to U, A_U is a self-adjoint operator on U. Furthermore, $U^{\perp}$ is an invariant subspace of A.*

Proof. We have $\langle A_U u, u' \rangle = \langle Au, u' \rangle = \langle u, Au' \rangle = \langle u, A_U u' \rangle$ for all $u, u' \in U$. Therefore A_U is a self-adjoint operator on U. Let $u \in U$ and $w \in U^{\perp}$, then we have $\langle u, Aw \rangle = \langle Au, w \rangle = 0$ always. Thus $Aw \in U^{\perp}$, and $U^{\perp}$ is invariant. ∎

The theorems of a self-adjoint operator on a finite dimensional inner product space is fruitful and interesting. We have,

Theorem 4.5.7 (Spectral Theorem for Self-Adjoint Linear Operator Defined over Finite Dimensional Vector Space): *Let the ground field be complex $\mathbb{C}$. Let A be a self-adjoint operator on a finite dimensional inner product space V. Then there is an*

orthonormal basis of V consisting of eigenvectors of A. Clearly, with respect to this basis, A is diagonal. We have $A = U^T \Delta U$, where U is a unitary matrix with column vectors forming the orthonormal basis, Δ a diagonal matrix with real entries. Furthermore, if A is diagonal with real values on the diagonal with respect to an orthonormal basis, then A is self-adjoint.

Proof. The characteristic equation of A is of degree n which is the dimension of the vector space V. Solving it, we find several eigenvalues. Call one of them λ_1, we know it is real. Let v_1 be an associated eigenvector which is non-zero. Hence we may assume it is a unit vector. We have the equation $Av_1 = \lambda v_1$. Let U be the subspace spanned by v_1. Then U is an invariant subspace of V. By the previous proposition, $W = U^\perp$ is an invariant subspace of V and A_W is self-adjoint on W. Instead of V, we work on W. We have an induction argument, and conclude that the matrix A is upper-triangular form with real terms on the diagonal. Since $A = \overline{A}^T$, then A must be of diagonal form with real terms on the diagonal with respect to the orthonormal basis. ∎

Corollary 4.5.8: *Let A be a real symmetric square matrix. Then A is diagonalizable by a real orthogonal matrix, i.e., there is a real orthogonal matrix O such that $O^T A O$ is diagonal.*

Proof. We follow the line of reasoning of the proof of the above theorem. First, we extend the scalar from the real number field $\mathbb{R}$ to the complex number field $\mathbb{C}$ by tensor product $V_{\mathbb{C}} = \mathbb{C} \otimes_{\mathbb{R}} V$. So we have a vector space $V_{\mathbb{C}}$ over the complex numbers $\mathbb{C}$, and A is hermitian or self-adjoint. Then we have a real eigenvalue λ with associated eigenvector v_1. We have v_1 as a complex vector. We may separate $v_1 = u_1 + iw_1$ into real and imaginary parts. Note that w_1 is a real vector. We have $Av_1 = Au_1 + iAw_1 = \lambda u_1 + i\lambda w_1$. We conclude that at least one of u_1, w_1 is non-zero and a real eigenvector associated with λ. We may assume the non-zero one is a unit real vector v_1'. As in the proof of the theorem, let U be the subspace spanned by v_1'. Then U is an invariant subspace of V. By the previous proposition, $W = U^\perp$ is an invariant subspace of V

and A_W is self-adjoint on W. Instead of V, we work on W. We have an induction argument, and conclude that the matrix A is upper-triangular form with real terms on the diagonal. Since $A = A^T$, then A must be of diagonal form with real terms on the diagonal with respect to the orthonormal matrix $O = [v_1, v_2, \ldots, v_n]$, i.e., $O^{-1}AO$ is diagonal. ∎

Example 4.14 (Classification of Quadratic Curves and Surfaces):

[Quadratic Curves]

In high school, we were taught that there are only three kinds of quadratic curves: parabola, hyperbola, and ellipse. Now we shall study the **claim**. Let us consider all possible quadratic equations as

$$ax^2 + bxy + cy^2 + dx + ey + f = 0.$$

Then the above equation can be written as

$$(x, y) \begin{pmatrix} a & b/2 \\ b/2 & c \end{pmatrix} \begin{pmatrix} x \\ y \end{pmatrix} + (x, y) \begin{pmatrix} d \\ e \end{pmatrix} + f = 0.$$

Let the 2×2 matrix be A. Then it follows from our Corollary 4.5.8 that the matrix A can be diagonalized by an orthogonal matrix D, i.e., $A = D\Lambda D^T$. Let $X = (x, y), E = (d, e)$, then the above equation can be rewritten as $XAX^T + XE^T + f = 0$. After we replace A by $D\Lambda D^T$, we have $XD\Lambda D^T X^T + XE^T + f = 0$. Rename $XD = Y$ and $X = YD^T$, we have $Y\Lambda Y^T + Y(D^T E^T) + f = 0$. We have only three possibilities: (1) one of the two eigenvalues is zero, the other one is non-zero, (2) both eigenvalues are non-zero, and they are of the same sign, (3) both eigenvalues are non-zero, and they are of different signs.

(1) One of the two eigenvalues is zero, the other one is non-zero. Without loss of generality we assume the equation is of the form

$$x^2 + dx + ey + f = (x + d/2)^2 + ey + f - d^2/4a = 0.$$

(a) If $e \neq 0$, then the curve is a parabola. (b) If $e = 0$ and $f - d^2/4 = 0$, then the curve is a coincident line. (c) If $e = 0$, and

$f - d^2/4 < 0$, then the curve is a pair of real parallel lines. (d) If $e = 0$, and $f - d^2/4 > 0$, then the curve is a pair of imaginary parallel lines.

(2) Both eigenvalues are non-zero, and they are of the same sign. We may assume that the equation is of the form,

$$x^2 + y^2 + dx + ey + f = (x + d/2)^2 + (y + e/2)^2 + f - d^2/4 + e^2/4 = 0.$$

(e) If $f - d^2/4 - e^2/4 = 0$, then the curve is a pair of two imaginary intersection lines. (f) If $f - d^2/4 - e^2/4 < 0$, then the curve is an ellipse. (g) If $f - d^2/4 - e^2/4 > 0$, then it is an imaginary ellipse.

(3) Both eigenvalues are non-zero, and they are of different signs. Without loss of generality we assume the equation is of the form,

$$x^2 - y^2 + dx + ey + f = (x + d/2)^2 - (y - e/2)^2 + f - d^2/4 + e^2/4 = 0.$$

(h) If $f - d^2/4 + e^2/4 = 0$, then the curve is a pair of two real intersected lines. (i) If $f - d^2/4 + e^2/4 < 0$, then it is a hyperbola. (j) If $f - d^2/4 + e^2/4 > 0$, then it is hyperbola.

In the above discussions, we find that if we exclude the degenerate cases (i.e., the quadratic curve degenerates into two straight lines) and the imaginary cases (we have to extend the real numbers to the complex numbers), we indeed have only ellipse, hyperbola and parabola.

[Quadratic Surfaces]

For a quadratic surface in $\mathbb{R}^3$, let us consider a general polynomial as

$$ax^2 + bxy + cxz + dyz + ey^2 + fz^2 + gx + hy + iz + j = 0.$$

Then the above equation can be written as

$$(x, y, z) \begin{pmatrix} a & b/2 & c/2 \\ b/2 & e & d/2 \\ c/2 & d/2 & f \end{pmatrix} \begin{pmatrix} x \\ y \\ z \end{pmatrix} + (x, y, z) \begin{pmatrix} g \\ h \\ i \end{pmatrix} + j = 0.$$

Let the first 3×3 matrix be A. Then it follows from our corollary that the matrix A can be diagonalized by an orthogonal matrix D, i.e., $A = D\Lambda D^T$. Let $X = (x, y, z)$, $E = (g, h, i)$, then the above equation can be rewritten as $XAX^T + XE^T + j = 0$. After we replace A by

DAD^T, we have $XDAD^TX^T + XE^T + j = 0$. Rename $XD = Y$ and $X = YD^T$, we have $YAY^T + Y(D^TE^T) + j = 0$. We have only five possibilities: (1) two of the three eigenvalues are zero, the other one is non-zero, (2) one of the three eigenvalues is zero and the other two eigenvalues are non-zero, the non-zero ones are of the same sign, (3) one of the three eigenvalues is zero and the other two eigenvalues are non-zero, the non-zero ones are of different signs, (4) all eigenvalues are non-zero, and they are of the same sign, (5) all eigenvalues are non-zero and they are of different signs. It is easy to deduce that two of them are of the same sign which is different from the other one.

After we dismiss the degenerate and imaginary cases, there are nine cases for quadratic surface as follows:

(1) (a) Parabola cylinder: with equation $x^2 + ry = 0$, $r \neq 0$; (2) (b) elliptic cylinder with equation $x^2/a^2 + y^2/b^2 = 1$, (c) elliptic paraboloid with equation $x^2/a^2 + y^2/b^2 = z$, (3) (d) hyperbolic paraboloid with equation $x^2/a^2 - y^2/b^2 = z$, (e) hyperbolic cylinder with equation $x^2/a^2 - y^2/b^2 = 1$, (4) (f) ellipsoid with equation $x^2/a^2 + y^2/b^2 + z^2/c^2 = 1$, (5) (g) elliptic cone with equation $x^2/a^2 + y^2/b^2 - z^2/c^2 = 0$, (h) hyperboloid of one sheet with equation $x^2/a^2 + y^2/b^2 - z^2/c^2 = 1$, (i) hyperboloid of two sheets with equation $x^2/a^2 + y^2/b^2 - z^2/c^2 = -1$. ∎

Example 4.15 (Classification of Partial Linear Differential Equations of Order 2): Let the following equation be given,

$$L = \sum_{i \leq j} a_{ij} \frac{\partial^2}{\partial x_i \partial x_j} + \text{lower order terms} = 0.$$

The *symbol* is the polynomial produced by replacing $\frac{\partial}{\partial x_i}$ by a variable y_i. The symbol of the above equation is $\sum_{ij} a_{ij} y_i y_j$. We consider the matrix of the quadratic form A as

$$A = \begin{pmatrix} a_{11} & (1/2)a_{12} & \cdots & (1/2)a_{1n} & \cdots \\ (1/2)a_{12} & a_{22} & \cdots & (1/2)a_{2n} & \cdots \\ \cdots & \cdots & \cdots & \cdots & \cdots \\ (1/2)a_{1n} & (1/2)a_{2n} & \cdots & \cdots & a_{nn} \end{pmatrix}.$$

We may select an orthonormal basis to make the above matrix diagonal. We shall thus assume the above matrix is already in diagonal form. We have the following classification of linear differential equations of order 2 as

(1) *Elliptic*: all eigenvalues are of the same sign.
(2) *Parabolic*: one of the eigenvalues is 0, the rest are of the same sign.
(3) *Hyperbolic*: the sign of one eigenvalue is different from all others.
(4) *Ultra-hyperbolic*: more than 1 positive, and more than 1 negative and no 0. ∎

Example 4.16 (Rayleigh[24] Quotient): For a hermitian complex matrix $M_{n \times n}$ and a non-zero vector v in $\mathbb{C}^n$, we define the Rayleigh quotient as

$$R(M, v) = \frac{v^* M v}{v^* v}.$$

In the real cases, *hermitian operator* becomes *symmetric*; $M^* = M^T$ and $v^* = v^T$. It is easy to see that $R(M, v) \in [\lambda_{\min}, \lambda_{\max}]$. It is obvious after observing that the Rayleigh quotient is the weighted average of eigenvalues of M:

$$R(M, v) = \frac{v^* M v}{v^* v} = \frac{\sum \lambda_i u_i^2}{\sum u_i^2}$$

where (λ_i, v_i) is the ith eigenpair after normalization and u_i is the ith coordinate of v with respect to the coordinate system $\{v_1, \ldots, v_n\}$. Recall that $v = \sum (v_i^* v) v_i$, thus $u_i = v_i^* v$.

Note that the above fact is untrue if M is not hermitian. Let

$$A = \begin{pmatrix} \epsilon & 1 \\ 0 & \epsilon \end{pmatrix}.$$

Then (ϵ, v_1) is the only eigenpair after normalization where $v_1 = [1, 0]^T$, and $R(A, v_1) = \epsilon$. Let $w = [1, 1]^T$, then $R(A, w) = 1/2 + \epsilon$ which is larger than ϵ.

---

[24]Rayleigh. English physicist and mathematician, 1842–1919, who with William Ramsay discovered argon which earned them the Nobel Prizes in 1904. His birth name was J. W. Strutt. In 1873, he inherited the title *Lord Rayleigh* from his father.

The above fact can be used to identify all eigenvalues of a hermitian matrix M. Let $\lambda_{max} = \lambda_1 \geq \lambda_2 \geq \cdots \geq \lambda_n = \lambda_{min}$ be the eigenvalues in decreasing order. Let V be the subspace which is perpendicular to v_1. Let N be the restriction of M to V. Then we have

$$\max_{v \in V}(R(N,v)) = \lambda_2.$$

Inductively we can find all λ_i, for $i = 3, \ldots, n$.

We shall consider the equation

$$R(M,v) = v^* M v, \quad \text{with } v^* v = 1$$

the restriction is the unit sphere in $\mathbb{C}^n$. Let us use *Lagrange multiplier* λ;

$$L(v) = v^T M v - \lambda(v^T v - 1).$$

We have

$$\frac{dL(v)}{dv} = 0$$

$$\Rightarrow 2Mv - 2\lambda v = 0$$

$$\Rightarrow Mv = \lambda v$$

and

$$R(m,v) = \frac{v^T M v}{v^T v} = \lambda \frac{v^T v}{v^T v} = \lambda.$$

Therefore the eigenvectors $v_1, v_2, \ldots, v_n$ of M are the critical points of the Rayleigh quotient and their eigenvalues $\lambda_1, \lambda_2, \ldots, \lambda_n$ are the stationary values of R.   ∎

Exercises

(1) Show the following *Polarization Identity*. Let V be an inner product space. Then $\forall v, u \in V$, we have

$$\langle v, u \rangle = (1/4)(|v + u|^2 - |v - u|^2 - i|v + iu|^2 + i|v - iu|^2).$$

(2) Given H_∞. We use the inner product $\langle v, u \rangle = \sum_i (1/2^i) a_i \bar{b_i}$. Let A be the shifting indices by 1 i.e., $A(a_1, a_2, a_3, \ldots) = (0, a_1, a_2, \ldots)$. Let $A^*(b_1, b_2, b_3, \ldots) = (1/2)(b_2, b_3, \ldots)$. Check that A^* is the adjoint of A.

(3) Let $\mathbb{C}^3$ be the inner product space with respect to the usual product. Find the adjoint of A where

$$A = \begin{bmatrix} 0 & 2 & i \\ i & 0 & 0 \\ 0 & 1 & 0 \end{bmatrix}.$$

(4) Let F_n be the nth Fourier matrix. Show that $((1/\sqrt{4})F_4)^4 = I$. Therefore the eigenvalues of $(1/\sqrt{4})F_4$ are $\{1, -1, i, -i\}$.

(5) Let V be a finite dimensional inner product space, we know the adjoint exists for any linear operator. Show that $(S + T)^* = S^* + T^*$, $(\alpha T)^* = \bar{\alpha} T^*$, $(T^*)^* = T$, $I^* = I$, $(ST)^* = T^* S^*$.

(6) Find the eigenvalues of the following matrix,

$$A = \begin{bmatrix} 2 & 1+i \\ 1-i & 3 \end{bmatrix}.$$

(7) Find the orthonormal basis such that the matrix A of the preceding exercise is of the diagonal form.

(8) A skew-hermitian form is a sesquilinear form $\phi : V \times V \mapsto \mathbb{C}$ such that

$$\phi(v, u) = -\overline{\phi(u, v)}.$$

Show that every skew-hermitian form can be written as i times a hermitian form.

(9) Show that over a finite dimensional vector space, the eigenvalues of a skew-hermitian matrix are all pure imaginary.

(10) For any $n \times m$ matrix A over $\mathbb{C}$. Define $A^* = \overline{A^T}$. Show that AA^* is self-adjoint, and A^*A is self-adjoint.

4.6 Minimal Problems and Positive-Definite Matrices. Many Variable Functions

Let A be a linear self-adjoint operator defined on an inner product space V with $\text{Dom}(A) = V$. We want to know when $\langle Ax, y \rangle$ is a new inner product $\langle\langle x, y\rangle\rangle = \langle Ax, y\rangle$ on V. As in the definition of an inner product, it must satisfy the three axioms of inner product: (1) $\langle\langle x, y\rangle\rangle = \langle Ax, y\rangle$ must be linear in x. This is satisfied by the linearity requirement and the domain requirement of A. (2) We must have $\langle\langle x, y\rangle\rangle = \langle Ax, y\rangle = \overline{\langle y, Ax\rangle} = \overline{\langle Ay, x\rangle} = \overline{\langle\langle y, x\rangle\rangle}$, this condition is satisfied iff A is self-adjoint. (3) $\langle\langle x, x\rangle\rangle = \langle Ax, x\rangle \geq 0$ and equal sign iff $x = 0$.

The third condition in the preceding paragraph is new. We shall define,

Definition A linear self-adjoint operator A with $\text{Dom}(A) = V$ is said to be *positive-definite* iff $\langle Ax, x\rangle \geq 0$, and equal sign iff $x = 0$. It is positive semi-definite iff $\langle Ax, x\rangle \geq 0$.   ∎

Remark: A self-adjoint linear operator A is positive-definite iff $\langle Ax, y\rangle$ is an inner product which will be written as $\langle x, y\rangle_A$.   ∎

We shall explain the meaning of positive-definite matrices in the real cases. Hence in the remaining part of this section, the ground field K is the real number field $\mathbb{R}$.

One of the problems in *Calculus* is to solve the local (isolated) minimal or maximal problems. At a point p, we want to decide locally around the point if a function $y = f(x_1, x_2, \ldots, x_n)$ has a maximal (profit) or minimal (cost) value. If we change the sign of f, then we just interchange the (isolated) maximal and (isolated) minimal point. Therefore, we shall only treat one of the two problems, (isolated) maximal or (isolated) minimal. Traditionally, we select the (isolated) minimal problem.

We shall further simplify the equation $y = f(x_1, x_2, \ldots, x_n)$ to an extent that we can handle. By moving the point p to the origin (replacing x_i by $x_i - a_i$), we may assume that we are discussing the situation around $(0, 0, \ldots, 0)$. By moving the graph $y = f(x_1, x_2, \ldots, x_n)$ up and down (replacing f by $(f - b)$), we may assume that the value is 0 at the origin $(0, 0, \ldots, 0)$, i.e., replacing x_i by 0 for all i, we have $y = 0$, the equation is reduced to $0 = f(0, 0, \ldots, 0)$. As usual, we first find all critical points, i.e., the points where the tangent plane is horizontal. In terms of *Calculus*, this means that partial derivatives $\frac{\partial f}{\partial x_i} = 0$ for all i at the origin $(0, \ldots, 0)$. Let us assume that the origin $(0, 0, \ldots, 0)$ is a critical point. In other words, the function f will have value 0 at the point $x = 0$ and with vanishing linear terms, we have

$$f(x_1, x_2, \ldots, x_n) = f_2(x_1, x_2, \ldots, x_n) + \cdots$$
$$= \sum_{i \leq j} b_{ij} x_i x_j + \text{higher terms.}$$

We shall assume that $f_2 \neq 0$, otherwise we will say the test fails. As long as the point $x = (x_1, x_2, \ldots, x_n)^T$ is close to the origin, the degree higher than 2 terms vanishes faster, and can be ignored. For simplicity, we shall rewrite f as

$$f = x^T A x + \text{higher terms}$$

where $x = (x_1, x_2, \ldots, x_n)^T$ and $A = (a_{ij})$ is an $n \times n$ matrix where

$$A = \begin{pmatrix} b_{11} & (1/2)b_{12} & \cdots & (1/2)b_{1n} \\ (1/2)b_{12} & b_{22} & \cdots & (1/2)b_{2n} \\ \cdots & \cdots & \cdots & \cdots \\ (1/2)b_{1n} & (1/2)b_{2n} & \cdots & b_{nn} \end{pmatrix}.$$

The local property is clearly determined by the matrix A (under the assumption that $A \neq 0$). We shall investigate the matrix A for the purpose of studying the (isolated) minimal problem. The matrix is real symmetric hence self-adjoint with respect to the usual inner product (i.e., the dot product) of the n-dimensional vector space $\mathbb{R}^n$. We have the following,

Proposition 4.6.1: *In the finite dimensional cases, both TT^* and T^*T are positive semi-definite.*

Proof. (1) $(TT^*)^* = TT^*$, $(T^*T)^* = T^*T$. (2) $\langle TT^*v, v \rangle = \langle T^*v, T^*v \rangle = |T^*v|^2 \geq 0$. $\langle T^*Tv, v \rangle = \langle Tv, Tv \rangle = |Tv|^2 \geq 0$. ∎

Example 4.17: Let V be a complete inner product space (i.e., Hilbert space). Let U be a closed subspace. Then $V = U \oplus U^\perp$. Let $W = U^\perp$. We have the projection P_U. We will show that P_U is positive semi-definite.

Let $v, v' \in V$ be any two vectors. Then $v = u+w, v' = u'+w'$ with $u, u' \in U, w, w' \in U^\perp$. We have $P_U(v) = u, P_U(v') = u'$. Moreover, we have $\langle P_U(v), v' \rangle = \langle u, u' + w' \rangle = \langle u, u' \rangle = \langle u, P_U(v') \rangle = \langle v, P_U(v') \rangle$. Therefore $P_U = P_U^*$. Furthermore, we have $\langle P_U(v), v \rangle = \langle u, u \rangle \geq 0$. We conclude that P_U is positive semi-definite. ∎

Example 4.18: The following 3×3 matrix is positive-definite,

$$A = \begin{pmatrix} 3 & 2 & 0 \\ 2 & 3 & 0 \\ 0 & 0 & 1 \end{pmatrix}$$

and the following 3×3 matrix is positive semi-definite

$$A = \begin{pmatrix} 3 & 2 & 0 \\ 2 & 3 & 0 \\ 0 & 0 & 0 \end{pmatrix}.$$ ∎

Remark: Let V be finite dimensional. A self-adjoint matrix A on V with $\text{Dom}(A) = V$ is positive semi-definite if $v^*Av \geq 0$ for all $v \in V$ (in the real case, a symmetric matrix A is said to be positive semi-definite if $v^T Av \geq 0$ for all $v \in V$). The self-adjoint matrix is said to be positive-definite if $v^*Av > 0$ for all $0 \neq v \in V$. (In the real case, a symmetric matrix A is said to be positive-definite if $v^T Av > 0$ for all $0 \neq v \in V$.) ∎

We have the following definition,

Definition Let $A = (a_{ij})$ be a real $n \times n$ matrix. The *principal minors* A_m for $m = 1, 2, \ldots, n$ is defined to be $A_m = \det(a_{ij})_{i,j \leq m}$. The *diagonal minors* are the determinants of the matrix after deleting a subset of rows and the same subset of columns. ∎

Example 4.19: Let A be defined as

$$A = \begin{pmatrix} 3 & 2 & 0 \\ 2 & 3 & 0 \\ 0 & 0 & 2 \end{pmatrix}.$$

Then $A_1 = 3, A_2 = 5, A_3 = 10$. ∎

We have the following proposition which connects the positive-definite matrix with the local (isolated) minimal problem.

Proposition 4.6.2: *Let the ground field be the real numbers* $\mathbb{R}$*. The function*

$$f(x) = x^T A x + \text{higher terms}$$

where $A \neq 0$ *is symmetric, has a local (isolated) minimal at the origin* $(0, 0, \ldots, 0)$ *iff the matrix* A *is positive-definite.*

Proof. For the local property, we shall consider an arbitrarily small neighborhood of the origin, hence we may ignore the higher degree terms of f and only look at the matrix A. Since A is symmetric $(A^T = A)$ and

$$f(x) = x^T A x + \text{higher terms} = \langle Ax, x \rangle + \text{higher terms}.$$

Then clearly $f(0) = 0$ is a (isolated) minimal iff $\langle Ax, x \rangle > 0$, iff A is positive-definite. ∎

So the criteria of a local (isolated) minimal of a function f becomes a criteria of the positive-definite of a matrix A. To further study the positive-definite property of a matrix, we recall the fundamental Gaussian row and column operations. Note that there are three kinds of row or column operations: (1) We interchange two rows or two columns. (2) We have linear operations of multiplying a row or a column by a non-zero constant. (3) We multiply a row by a constant and add the result to a different row or multiply a column by a constant and add the result to a different column.

Definition A symmetric matrix A can be *straightly reduced to a diagonal form* if LAL^T is diagonal where L is the multiple of a sequence of elementary matrices of the third kind which corresponds to a simple procedure of clearing the lower half of the matrix A. ∎

Example 4.20: The following matrix can be straightly reduced to a diagonal form,

$$A = \begin{pmatrix} 3 & 2 & 0 \\ 2 & 3 & 0 \\ 0 & 0 & 1 \end{pmatrix}.$$

The following matrix cannot be straightly reduced to a diagonal form,

$$A = \begin{pmatrix} 0 & 2 & 0 \\ 2 & 3 & 0 \\ 0 & 0 & 1 \end{pmatrix}. \qquad \blacksquare$$

Proposition 4.6.3: *Let us consider $\mathbb{R}^n$ and let A be an $n \times n$ symmetric matrix. The following four statements are equivalent: (1) A is positive-definite. (2) All eigenvalues λ_i of it are positive. (3) All principal minors $A_m > 0$ for $m = 1, 2, \ldots, n$. (4) A can be straightly reduced to a diagonal form with all the pivots δ_i positive.*

Proof. We shall show that (1) $\Longleftrightarrow$ (2). Let us assume (1). Let $v_1, v_2, \ldots, v_n$ be an orthonormal basis such that $A = C\Lambda C^T$, where v_i is the ith column vector of C. Let $v = \sum_i a_i v_i$. Then $Av = \sum_i \lambda_i a_i v_i$ and $\langle Av, v \rangle = \sum_i \lambda_i |a_i|^2$. Therefore $\lambda_i > 0$ for all i. Let us assume (2). With the same notation as above, we have $\langle Av, v \rangle > 0$ for all $0 \neq v \in V$.

Let us show that (1) (or (2)) $\Longrightarrow$ (3). From the theory of determinant and eigenvalues we have determinant$(A) = \prod_i \lambda_i = A_n > 0$. So we have $A_n > 0$. For any $1 \leq m \leq n$, we may consider the subspace U generated by $v_1, \ldots, v_m$. Then the restriction of A to U is positive-definite, and $\prod_{i=1}^m \lambda_i = A_m$ is positive.

Let us show that (3) $\Longrightarrow$ (4). Since $A_1 > 0$ we have $a_{11} > 0$. Then we may use it to clean all the terms in the first row and the first column. Note that it is a sequence of elementary operations of the third kind. Then clearly we have $A_2 = a_{11}a'_{22}$, where a'_{22} is the term after the cleaning we just mentioned. We have $a'_{22} > 0$. Note that the process for cleaning the lower half of A is the same as the cleaning process of the lower half of the smaller $m \times m$ submatrix at the left corner. Therefore $A_m = \prod_{i=1}^m \delta_i > 0$. From those inequalities, we deduce that the matrix can be straightly reduced to diagonal form and $\delta_i > 0$ for all i.

Let us show that (4) $\implies$ (1). Let us compute $\langle Av, v \rangle = \langle L\Sigma L^T v, v \rangle = v^T L\Sigma L^T L^T v = (L^T v)^T \Sigma (L^T v)$. Let us take $L^T v = u$. Note that L is an isomorphism. So u can be any vector $u_1, u_2, \ldots, u_n$. Therefore we conclude that $\langle Av, v \rangle = \sum_i \delta_i |u_i|^2 > 0$ for any $0 \neq v \in \mathbb{R}^n$. ∎

We will give the following useful corollaries,

Corollary 4.6.4: *Let us consider a real function $f = x^T A x +$ higher terms, where A is real symmetric. Then $z = f$ has a local (isolated) minimal iff $A_m > 0$ for $m = 1, 2, \ldots, n$.*

Proof. It follows from Proposition 4.6.3. ∎

Corollary 4.6.5: *Let us consider a real function $f = x^T A x +$ higher terms, where A is real symmetric. Then $z = f$ has a local (isolated) minimal iff A can be straightly reduced to a diagonal form with all the pivots δ_i positive.*

Proof. It follows from Proposition 4.6.3. ∎

Proposition 4.6.6: *Let us consider a real symmetric matrix A. Then A is positive-definite iff $A = R^T R$ where R is an $n \times n$ matrix with the column vectors a basis of R^n, i.e., R is a non-singular matrix.*

Proof. ($\impliedby$) It follows that $0 = \langle Av, v \rangle = (Av)^T v = v^T A^T v = v^T R^T R v = (Rv)^T Rv = |Rv|^2$ hence $Rv = 0$ and $v = 0$.

($\implies$) It follows that $A = LDL^T$ where D is a diagonal matrix with positive number on the diagonal. Let $\sqrt{D}$ be the positive diagonal matrix with $(\sqrt{D})^2 = D$ and $R^T = L\sqrt{D}$. Then it is easy to see $A = R^T R$. ∎

Example 4.21 (Second Derivative Test): Let us study the origin for a function $y = f(x_1, x_2, \ldots, x_n)$. After we move $y, x_1, x_2, \ldots, x_n$ linearly, we may assume that $f(0, 0, \ldots, 0) = 0$ and the origin is a critical point, i.e., $\partial f/\partial x_i = 0$ at $x = 0$ for all i. We may write $f(x_1, x_2, \ldots, x_n) = \sum_{i \geq j} b_{ij} x_i x_j +$ higher terms. Let $A = (a_{ij})$ be as the above, and $D(x_1, x_2, \ldots, x_n) = (\partial^2 f/\partial x_i \partial x_j)$. Then we see that $D(0) = 2A$. As usual we assume $A \neq 0$. Clearly A is positive-definite iff $D(0)$ is positive-definite. Let $2A_m(0) = \det(\partial^2 f/\partial x_i \partial x_j)_{i,j \leq m}(0)$ for $m = 1, 2, \ldots, n$. The second derivative test for minimum is $D_m(0) > 0$ for $m = 1, 2, \ldots, n$. ∎

Example 4.22 (Second Derivative Test for Small Number of Variables): We shall use the assumptions of the previous example. If there is only one variable. Let the variable be x. Then $D_1(x) = (f''(x))$ and the theory claims that $f'(0) = 0, f''(0) > 0$ is the criteria.

If there are two variables. Let the variables be x, y for a function $f(x, y)$. Let us assume that $f(0,0) = 0$ and the origin is a critical point, i.e., $f_x(0,0) = 0$, $f_y(0,0) = 0$. Then we have

$$D_1(x, y) = (f_{xx}), \quad D_2(x, y) = \begin{pmatrix} f_{xx} & f_{xy} \\ f_{yx} & f_{yy} \end{pmatrix}.$$

The theory claims that the origin is a (isolated) minimal point iff $\det(D_1)(0) = f_{xx}(0) > 0$ and $\det(D_2)(0) = (f_{xx}f_{yy} - f_{xy}^2)(0) > 0$ which is precisely what we call the second derivative test for functions of two variables in multivariable *Calculus*.

Let us consider the cases of three variables x, y, z for a function $f(x, y, z)$. Let us assume that $f(0,0,0) = 0$ and the origin is a critical point. Let us assume that the following $D|_0 \neq 0$,

$$D = \begin{pmatrix} f_{xx} & f_{xy} & f_{xz} \\ f_{yx} & f_{yy} & f_{yz} \\ f_{zx} & f_{zy} & f_{zz} \end{pmatrix}.$$

The theory claims that the origin is a (isolated) minimal point iff $f_x(0) = 0$, $f_y(0) = 0$, $f_z(0) = 0$ and $f_{xx}(0) > 0$, $D_2(0) = (f_{xx}f_{yy} - f_{xy}^2)(0) > 0$ and $D_3(0) = (f_{xx}f_{yy}f_{zz} + f_{xy}f_{yz}f_{zx} + f_{xz}f_{xy}f_{yz} - f_{xz}f_{yy}f_{xz} - f_{xy}^2 f_{zz} - f_{xx}f_{yz}^2)(0) > 0$ which is the second derivative test for functions of three variables. ∎

Example 4.23: Let us study the following equation to decide if it has a local minimum at the origin,

$$f(x, y, z) = 3x^2 + 6xy + 2xz + 4y^2 + 3yz + 2z^2$$

we shall rewrite the above equal as $v^T A v$ with $v = (x, y, z)^T$ and A as

$$A = \begin{pmatrix} 3 & 3 & 1 \\ 3 & 4 & 3/2 \\ 1 & 3/2 & 0 \end{pmatrix}.$$

Then we have $A_1 = 3$, $A_2 = 3$, $A_3 = -7/4$. Therefore the origin is not a (isolated) minimal point. ∎

Example 4.24: Let A be a real symmetric positive-definite matrix. Let us consider the following quadratic equation,

$$f(x) = \frac{1}{2}x^T A x - x^T b.$$

We may want to find its local minimums. Following *Calculus*, geometrically, the positive-definite of a matrix is associated with the n-dimensional ellipsoid. We have the *gradient* at a point x to be

$$f'(x) = \begin{pmatrix} \dfrac{\partial f(x)}{\partial x_1} \\[1mm] \dfrac{\partial f(x)}{\partial x_2} \\[1mm] \cdots \\[1mm] \dfrac{\partial f(x)}{\partial x_n} \end{pmatrix} = \frac{1}{2}A^T x + \frac{1}{2}Ax - b = Ax - b.$$

The minimal point of $f(x)$ is among the solutions of $Ax = b$. On the other hand, if $Ax = b$, then we have

$$f(y) - f(x) = \frac{1}{2}y^T A y - y^T b - \frac{1}{2}x^T A x + x^T b$$

$$= \frac{1}{2}y^T A y - y^T A x + \frac{1}{2}x^T A x \qquad (\text{set } b = Ax)$$

$$= \frac{1}{2}(y - x)^T A(y - x) > 0 \quad \text{if } y \neq x.$$

Since A is positive-definite, the above term > 0 if $y \neq x$. Therefore any solution of $Ax - b = 0$ is a minimal point of $f(x)$. Hence we show the minimum problem of $x^T A x - x^T b$ and the problem of solving $Ax = b$ are equivalent.

The minimal principle happens very often in science and engineering. The oldest one was the law of refraction of Fermat.[25] He computed the minimal time for the light ray to travel from one point to another point in a different medium. Later on in physics, there is the principle of least *action* which is energy × time. In a mixture of liquids, the heavy ones sink to the bottom to minimize their potential energy. There are many examples in engineering. If the equation happens to be a quadratic as above, then it is reduced to solve a linear equation.

On the other hand, if the linear equations are too big to be computed, then we may turn to the table, and convert it to a minimal problem of successively approximating the minimal point of a system of quadratic equations. This is the kernel of the popular method of the *Conjugate Gradient Method* (cf. Section 6.4).　　　■

Proposition 4.6.7: *Let A be a linear operator on $\mathbb{R}^n$. Suppose that A is symmetric and positive-definite. Then the image of an $(n-1)$-dimensional sphere is an $(n-1)$-dimensional ellipsoid.*

Proof. Left to the reader.　　　　　　　　　　　　　　　■

We have the following theorem for positive semi-definite,

Proposition 4.6.8: *Let us consider $\mathbb{R}^n$ and let A be an $n \times n$ symmetric matrix. The following five statements are equivalent:* (1) *A is positive semi-definite,* (2) *all eigenvalues λ_i of it are non-negative,* (3) *all principal minors $A_m \geq 0$ for $m = 1, 2, \ldots, n$,* (4) *A can be straightly reduced to a diagonal form with all the pivots δ_i non-negative,* (5) *A can be written as $R^T R$, where R may be singular.*

---

[25] Fermat, P. French lawyer and mathematician. 1607–1665. He was famous for Fermat's last problem.

[Max–Min Principle]

It started when Fermat deduced the *law of refraction* of the light ray from the minimization of the *time* of traveling of a ray of light as follows,

Law of Refraction

Suppose that there are two different mediums, say air and water, and the light ray traveling from one point P to a point Q in a different medium. Let the speeds of light in the two mediums be v_1, v_2 respectively as in the following diagram,

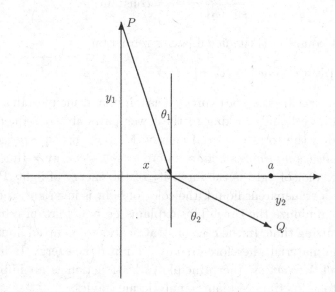

where the point P is with coordinate $(0, y_1)$ and the coordinate of Q is $(a, -y_2)$. The speed of light in the medium above the horizontal line is v_1 and the speed of the light in the medium below the horizontal line is v_2. Then the traveling light for the light ray through the point x is

$$t = f(x) = \frac{\sqrt{x^2 + y_1^2}}{v_1} + \frac{\sqrt{(a - x)^2 + y_2^2}}{v_2}.$$

The necessary condition for x to be a minimal point of $f(x)$ is $f'(x) = 0$ which is explicitly,

$$f'(x) = \frac{x}{v_1 \sqrt{y_1^2 + x^2}} - \frac{a - x}{v_2 \sqrt{(y_2^2 + (a - x)^2)}} = 0.$$

Recall that

$$\frac{x}{\sqrt{y_1^2 + x^2}} = \sin(\theta_1) \qquad \frac{a - x}{\sqrt{y_2^2 + (a - x)^2}} = \sin(\theta_2).$$

The above equation can be written as

$$\frac{\sin(\theta_1)}{\sin(\theta_2)} = \frac{v_1}{v_2} = \text{constant}.$$

The last equation is the usual law of refraction.

[Principle of Least Action]

Many natural laws can be expressed as minimal principles. In a mixed liquid, heavy liquids sinking to the lower parts are consequences of minimizing the total potential energy. Maupertuis[26] proposed that *nature always minimizes action*, where *action* is energy × time. Note that the color of light is the intensity of the energy of light. Passing through a transparent liquid, the color of light is invariant, therefore we only minimize the time. This explains Fermat's law of refraction of minimizing time. In the case of stratification of a mixed liquid, the time is immaterial, therefore we only minimize the energy. In modern *Quantum Mechanics*, the principle of least action is an important cornerstone for Hamiltonian[27] analytic mechanics.

[Max–Min Principle]

In this subsection, we assume that the matrix A is hermitian, therefore it has all real eigenvalues $\lambda_1 \leq \lambda_2 \leq \cdots \leq \lambda_n$. We have the following definition,

---

[26] Maupertuis, P. French mathematician. 1698–1759.
[27] Hamilton, W. Irish mathematician. 1805–1865.

Definition We define the Rayleigh–Ritz[28] quotient $R_A : C^n \setminus (0) \mapsto \mathbb{R}$ defined as

$$R_A(x) = \frac{\langle Ax, x \rangle}{\langle x, x \rangle}.$$

■

We have the following theorem,

Theorem 4.6.9: *Let A be hermitian. We have*

$$\lambda_k = \min_U \{ \max_x \{ R_A(x) : x \in U, |x| = 1 \} : \dim(U) = k \} \qquad (1)$$

and

$$\lambda_k = \max_U \{ \min_x \{ R_A(x) : x \in U, |x| = 1 \} : \dim(U) = n - k + 1 \}.$$

$$(2)$$

In particular,

$$\lambda_1 \le R_A(x) \le \lambda_n \quad \forall |x| = 1.$$

Proof. Since the matrix A is hermitian, it can be diagonalized, and there is an orthonormal basis $\{u_1, \ldots, u_n\}$ with $\langle u_i, u_j \rangle = \delta_{ij}$. Let U be any vector subspace of dimension k, and let $V_k = \langle u_k, \ldots, u_n \rangle$, a vector subspace of dimension $n - k + 1$. The intersection of U and V_k is at least 1-dimensional. Therefore there is an element $0 \ne u \in U$ such that

$$u = \sum_k^n a_i u_i$$

and whose Rayleigh–Ritz quotient is

$$R_A(u) = \frac{\sum_k^n \lambda_i |a_i|^2}{\sum_k^n |a_i|^2} \ge \lambda_k$$

hence we have

$$\max\{ R_A(x) : x \in U, |u| = 1 \} \ge \lambda_k.$$

Since the above equation is true for all U, we have

$$\lambda_k \le \min_U \{ \max_x \{ R_A(x) : x \in U, |x| = 1 \} : \dim(U) = k \}.$$

---

[28]Ritz, W. Swiss physicist. 1878–1909.

To prove the other inequality, let $U = \langle u_1, \ldots, u_k \rangle$. Then $\dim(U) = k$, and

$$\max\{R_A(x) : x \in U \text{ and } |u| = 1\} \leq \lambda_k.$$

We finish the proof of Eq. (1). The proof of Eq. (2) is similar by considering $V'_k = \langle u_1, \ldots, u_k \rangle$.   ∎

Example 4.25: Let A be the following matrix which is not hermitian,

$$A = \begin{bmatrix} 0 & 1 \\ 0 & 0 \end{bmatrix}.$$

We have $\lambda_1 = \lambda_2 = 0$. Let us use formula (1) to compute λ_2. We have

$$\min_U \{\max_x \{R_A(x) : x \in U, |x| = 1\} : \dim(U) = 2\}.$$

Let $x = \begin{bmatrix} \cos\theta \\ \sin\theta \end{bmatrix}$, then we have $Ax = \begin{bmatrix} \sin\theta \\ 0 \end{bmatrix}$ and we have

$$\min_U \{\max_x \{R_A(x) : x \in U, |x| = 1\} : \dim(U) = 2\} > 0 = \lambda_1 = \lambda_2.$$

∎

Exercises

(1) Determine if the following equation has a local isolated minimal at the origin of $\mathbb{R}^3$,

$$f(x, y, z) = x^2 + xy + xz + y^2 + yz + z^2.$$

(2) Show that $f(x, y, z) = b_{11}x^2 + b_{12}xy + b_{13}xz + b_{22}y^2 + b_{23}yz + b_{33}z^2$ has a (isolated) minimal at the origin implies that $b_{33} > 0$.

(3) Find the range of a such that the following matrix A is positive-definite,

$$A = \begin{pmatrix} a & 1 & 1 \\ 1 & a & 1 \\ 1 & 1 & a \end{pmatrix}.$$

(4) Let A be real symmetric. Show that if A is positive-definite, then so is A^m for $m > 1$ and A^{-m} for $m \geq 1$.

(5) If A, B are real symmetric and positive-definite, then $A + B$ is real symmetric and positive-definite.

(6) Let

$$A = \begin{pmatrix} 2 & 1 & 1 \\ 1 & 2 & 1 \\ 1 & 1 & 2 \end{pmatrix}.$$

Find a non-singular matrix R such that $A = R^T R$.

(7) Suppose that a matrix A is symmetric, let $f(x) = x^T A x - x^T b$. Prove that $\text{grad}(f) = \triangledown f(x) = f'(x)$, and

$$f'(x) = \begin{pmatrix} \dfrac{\partial f(x)}{\partial x_1} \\ \dfrac{\partial f(x)}{\partial x_2} \\ \cdots \\ \dfrac{\partial f(x)}{\partial x_n} \end{pmatrix} = \frac{1}{2} A^T x + \frac{1}{2} A x - b = A x - b.$$

(8) Let us use the notations in the Rayleigh quotient. Prove that $\lambda_n = \max\{R(x) : |x| = 1\}$.

(9) Let us use the notations in the Rayleigh quotient. Prove that $\lambda_1 = \min\{R_a(x) : |x| = 1\}$.

(10) Find the minimal values in $\mathbb{R}^2 \backslash (0)$ of the following function,

$$R(x, y) = \frac{x^2 + xy + y^2}{2x^2 + 3y^2}.$$

(11) Prove Proposition 4.6.7.

(12) Prove Proposition 4.6.8.

4.7 Unitary Operator

Let the ground field be the real field $\mathbb{R}$ or the complex field $\mathbb{C}$. One of the important properties of an inner product space V is that it is a metric space. Naturally we want to study the *isometries* of this space. We have the following definition,

Definition An operator A with $\text{Dom}(A) = V$ is said to be an *isometry* iff

$$\langle Av, Au \rangle = \langle v, u \rangle, \quad \forall v, u \in V. \qquad \blacksquare$$

Example 4.26: A rotation in $\mathbb{R}^n$ is an isometry and a map A of $\ell^2\{(a_1, a_2, \ldots) : \sum |a_j|^2 < \infty\}$ which sends $(a_1, a_2, \ldots) \mapsto (0, a_1, a_2, \ldots)$ is an isometry. Note that an isometry will preserve the lengths and angles. $\qquad \blacksquare$

Proposition 4.7.1: *An isometry A is injective.*

Proof. Suppose $Av = Aw$. Then $A(v - w) = 0$ and $\langle A(v - w), A(v - w) \rangle = |v - w|^2 = 0$. We conclude that $v = w$. $\qquad \blacksquare$

Definition An operator A is said to be a *unitary operator* iff: (1) it is an isometry, and (2) it is an onto map. $\qquad \blacksquare$

Example 4.27: In the finite dimensional cases, we have for $n \times n$ matrices A, B, if $AB = I$, then we always have $BA = I$ (cf. Chapter 2). In $\ell^2 = \{(a_1, a_2, \ldots) : a_i \in C, \sum_i |a_i|^2 < \infty\}$, let A be defined as $A(a_1, a_2, \ldots) = (0, a_1, a_2, \ldots)$. It is easy to see that A^* defined by $A^*(b_1, b_2, \ldots) = (b_2, b_3, \ldots)$ only satisfies $A^*A = I$, while $AA^* \neq I$. In fact, A is not an *onto* map. $\qquad \blacksquare$

Proposition 4.7.2: *Let a linear operator A have an adjoint A^*. Then we have: (1) the operator A is a unitary operator, iff (2) A^{-1} exists and $A^* = A^{-1}$, iff (3) $AA^* = A^*A = I$.*

Proof. (1) $\Longrightarrow$ (2): Since A is injective and surjective, so the inverse of A exists and is a linear operator. We always have $\langle Av, Au \rangle = \langle v, A^*Au \rangle = \langle v, u \rangle$ for all $v, u \in V$. Therefore we have $\langle v, (A^*A - I)u \rangle = 0$. Let $v = (A^*A - I)u$. We conclude that $A^*A - I = 0$. In other words, $A^* = A^{-1}$.

(2) $\Longrightarrow$ (1): Certainly A is onto and A^* exists. Then we have $\langle Av, Au \rangle = \langle v, A^*Au \rangle = \langle v, Iu \rangle = \langle v, u \rangle$.

(2) $\Longleftrightarrow$ (3): Easy. $\qquad \blacksquare$

Remark: In the finite dimensional cases, isometry implies onto. In fact, by considering the dimensions, it follows from the *dimension theory of vector spaces* that an injective linear operator must be surjective for finite dimensional cases. Clearly, an isometry is injective, hence it must be surjective. Usually, in the finite dimensional case, we treat *unitary* as *isometry*. ∎

Proposition 4.7.3: *All unitary linear operators form a group.*

Proof. It is easy. ∎

Example 4.28 (Unitary Geometry): The Erlanger program of Felix Klein[29] claims that every geometry is the study of invariants under some group. We have the *unitary group* consisting of all unitary maps of an inner product space. Therefore we have a unitary geometry associated with the inner product. There we have the concepts of length, angle, polygon, polyhedron, etc. The reader is referred to H. Weyl's[30] "The Theory of Groups and Quantum Mechanics", Chapter 1, Unitary Geometry. ∎

In the real finite dimensional cases, a linear operator A is unitary iff it is orthogonal.

We have the following proposition about the places of the eigenvalues of a unitary operator.

Proposition 4.7.4: *Given a unitary operator A with an adjoint A^*. Let λ be an eigenvalue of A, then $|\lambda| = 1$.*

Proof. Let $Av = \lambda v$ with $v \neq 0$. Then we have $A^* = A^{-1}$ and $A^*v = A^{-1}v = (1/\lambda)v$, and $\langle v, A^*v \rangle = \langle Av, v \rangle = \lambda \langle v, v \rangle$. Hence $\bar{\lambda}\lambda = 1$ which means $|\lambda|^2 = 1$. Since $|\lambda| \geq 0$, we must have $|\lambda| = 1$. ∎

In the finite dimensional case, we have the *Schur*[31] *decomposition theorem*, or traditionally it is named as *Schur's lemma*,

---

[29] Klein, F. German mathematician. 1849–1925.
[30] Weyl, H. German mathematician, physicist and philosopher. 1885–1955.
[31] Schur, I. Russian mathematician. 1875–1941.

Theorem 4.7.5 (The Schur Decomposition Theorem): *Let V be a finite dimensional complex inner product space and A be any square matrix, then there exists a unitary matrix B such that $BAB^* = C$ an upper-triangular matrix. (In other words, let T be the linear operator whose matrix form is A, then there exists an orthonormal basis $v_1, v_2, \ldots, v_n$, such that the matrix form of T is upper-triangular.)*

Proof. View this square matrix as the matrix form of a linear operator T. The linear operator has an eigenvalue λ and an associated vector $v \neq 0$. We may assume v is a unit, and complete it to an orthonormal basis $v, u_2, \ldots, u_n$. Let D be its matrix with respect to the orthonormal basis $v, u_2, \ldots, u_n$. Then D is of the form,

$$D = \begin{bmatrix} \lambda & a_{12} & a_{13} & * & a_{1n} \\ 0 & a_{22} & a_{23} & * & a_{2n} \\ 0 & a_{32} & a_{33} & * & a_{3n} \\ * & * & * & * & * \\ 0 & a_{n2} & a_{n3} & * & a_{nn} \end{bmatrix}.$$

Let us denote the $(n-1) \times (n-1)$ submatrix right-lower corner by F. Then F is a linear operator defined on the subspace spanned by $u_2, u_3, \ldots, u_n$. We may use the mathematical induction on the dimension to find an orthonormal basis $v_2, v_3, \ldots, v_n$ such that F will be upper-triangular. It is then easy to see that T is upper-triangular with respect to $v, v_2, v_3, \ldots, v_n$. ∎

Remark: We have shown in Chapter 2 that any monic polynomial $f(\lambda)$ can be realized as the characteristic polynomial of a square matrix. On the other hand, the eigenvalues of a triangle matrix are the diagonal terms of it. So to find the roots of a general polynomial is just to triangularize a matrix. The above theorem seems to have provided a way to solve any polynomial equation! Stop here. It is a tautology. Because in the proof of the theorem, we assume that we can find an eigenvalue. The QR method comes to rescue us. Later on in Section 6.3, we will explain how to use the QR method to approximately find the triangle form of a square matrix. ∎

Exercises

(1) For what values of α is the following matrix isometry,

$$A = \begin{bmatrix} \alpha & 0 \\ 1 & 1 \end{bmatrix}?$$

(2) For what values of α is the following matrix isometry,

$$A = \begin{bmatrix} \alpha & 1/2 \\ 1/2 & \alpha \end{bmatrix}?$$

(3) Show that every orthogonal matrix ($AA^T = I$) over $\mathbb{R}$ is unitary.

(4) Let $\pi : \mathbb{Z} \mapsto \mathbb{Z}$ be a one-one onto map. Then it induces a map on ℓ^2 indexed by $\mathbb{Z}$, where $\ell^2 = \{\ldots, a_{-2}, a_{-1}, a_0, a_1, a_2, \ldots :$ $\sum_{i=-\infty}^{i=\infty} |a_i|^2 < \infty\}$. We still name it as π. Show that π has an adjoint.

(5) Find a unitary matrix U such that UAU^* is an upper-triangular matrix with the matrix A defined as

$$A = \begin{bmatrix} 1 & 1 & 3 \\ 0 & 2 & 2 \\ 0 & 3 & 1 \end{bmatrix}.$$

(6) Show that given a reflection A of $\mathbb{R}^3$, -1 is an eigenvalue of A.

(7) Find a linear operator T, such that T^* exists, and $T^*T = I$ while $TT^* \neq I$.

(8) Show that for finite dimensional case, $T^*T = I$ implies $TT^* = I$.

4.8 Spectrum Theorem

Since the beginning of *Quantum Mechanics* (Heisenberg's matrix mechanics and Dirac–von Neumann's Hilbert space formulation of *Quantum Mechanics*), the concepts of *eigenvalues* and *eigenvectors* assume prominent positions and were matched with the physical ideas of spectrum. The spectral theorem provides conditions under which an operator or a matrix can be diagonalized (i.e., represented as a diagonal form in some basis). Those problems are known in the infinite dimensional case, and the theory is complicated. In this section we will emphasize the finite dimensional cases.

We have the following definition,

Definition A linear operator A is *bounded by a number L* iff $|Ax| \leq L|x|$ for all $x \in \text{Dom}(A)$. A linear operator A is *bounded* iff it is bounded by some number L. ∎

Definition A bounded linear operator T of an inner product space V is said to be *normal* iff: (1) T has an adjoint T^*, (2) $TT^* = T^*T$. ∎

Example 4.29: In the finite dimensional cases, the self-adjoint operators $(A = A^*)$, the skew-hermitian matrices $(A^* = -A)$, and unitary operators $(A^*$ always exists, and $AA^* = A^*A = I)$ are normal operators. ∎

Example 4.30: There is a major difference between real field $\mathbb{R}$ and complex field $\mathbb{C}$. We shall prove later that over complex numbers $\mathbb{C}$, every normal matrix can be diagonalized. Over the real field $\mathbb{R}$, a normal matrix may not be *diagonalizable*. Say a real matrix A with

$$A = \begin{bmatrix} 0 & 1 \\ -1 & 0 \end{bmatrix}.$$

Then A is skew-hermitian, hence it is normal. Clearly it cannot be diagonalized, because it has no real eigenvalue. ∎

Proposition 4.8.1: *Let T be a linear operator on a finite dimensional inner product space V. Then T is normal iff $|Tv| = |T^*v|$ for all $v \in V$.*

Proof. In the finite dimensional case, we know the existence of T^*. Then we have

$$T \text{ is normal} \iff T^*T - TT^* = 0$$

$$\iff \langle (T^*T - TT^*)v, v \rangle = 0 \ \forall v \in V$$

$$\Longleftrightarrow \langle T^*Tv, v \rangle = \langle TT^*v, v \rangle \ \forall v \in V$$

$$\Longleftrightarrow |Tv|^2 = |T^*v|^2 \ \forall v \in V. \qquad \blacksquare$$

Corollary 4.8.2: *Let T be a normal linear operator of an inner product space. Then we have*

(1) $nullspace(T) = nullspace(T^*)$.
(2) *If λ is an eigenvalue of T, then $\overline{\lambda}$ is an eigenvalue of T^* and they have the same associated eigenvectors.*
(3) *If λ_1, λ_2 are two distinct eigenvalues of A with associated eigenvectors v_1, v_2 respectively, then we have $\langle v_1, v_2 \rangle = 0$.*

Proof.

(1) It follows from the proposition and the positive-definite of the inner product, we have $\forall v \in \ker(A)$ $\langle Av, Av \rangle = 0 \Longleftrightarrow \langle v, A^*Av \rangle = 0 \Longleftrightarrow \langle v, AA^*v \rangle = 0 \Longleftrightarrow \langle A^*v, A^*v \rangle = 0 \Longleftrightarrow v \in \ker(A^*)$.
(2) First we observe that T is normal iff $T - \lambda I$, $(T - \lambda I)^* = T^* - \overline{\lambda} I$ are normal. We conclude that $|(T - \lambda I)v| = |(T^* - \overline{\lambda} I)v|$.
(3) We have

$$(\lambda_1 - \lambda_2)\langle v_1, v_2 \rangle = \langle \lambda_1 v_1, v_2 \rangle - \langle v_1, \overline{\lambda_2} v_2 \rangle$$

$$= \langle Tv_1, v_2 \rangle - \langle v_1, T^*v_2 \rangle = 0.$$

Since $\lambda_1 - \lambda_2 \neq 0$, then we must have $\langle v_1, v_2 \rangle = 0$. $\qquad \blacksquare$

The theory of normal operators in general is complicated, we shall restrict only to the finite dimensional cases. We have,

Theorem 4.8.3 (Spectral Theorem for Normal Operator on a Finite Dimensional Complex Space): *Let V be a finite dimensional complex vector space and T is a linear operator. Then T is normal iff there is an orthonormal basis of V consisting of eigenvectors of T.*

Proof. ($\Longleftarrow$) It is clear.

($\Longrightarrow$) It follows from Schur's theorem that with respect to an orthonormal basis, the matrix form A of T is upper-triangular. We deduce the theorem from $AA^* = A^*A$. Let A be the following matrix,

$$A = \begin{bmatrix} a_{11} & a_{12} & a_{13} & * & a_{1n} \\ 0 & a_{22} & a_{23} & * & a_{2n} \\ 0 & 0 & a_{33} & * & a_{3n} \\ * & * & * & * & * \\ 0 & 0 & 0 & * & a_{nn} \end{bmatrix}.$$

By direct computation, we have the (11) position of A^*A is $|a_{11}|^2$, while the (11) position of AA^* is $\sum_i |a_{1i}|^2$. For them to be equal, we must have $a_{1i} = 0$, $\forall i \geq 2$. After this we shall consider the (22) position. It is routine to conclude that $a_{2i} = 0, \forall i \geq 3$. After n steps, we conclude that A must be diagonal. Our theorem follows. ∎

In the finite dimensional cases, equivalently, we may write A as a linear combination of pairwise orthogonal projections, and call it *spectral decomposition*. Let

$$V_\lambda = \{v \in V : Av = \lambda v\}.$$

Note that v may be 0 vector in the above definition, and we call it the eigenspace associated with λ (it is the set of all eigenvectors associated with λ and the 0 vector). Let P_λ be the orthogonal projection onto V_λ. Then we have

Proposition 4.8.4: *With the notations above, we have*

$$A = \lambda_1 P_{\lambda_1} + \cdots + \lambda_m P_{\lambda_m}$$

where $\lambda_1, \ldots, \lambda_m$ form the complete set of all distinct eigenvalues of A

Proof. Routine. ∎

Recall the Lagrange interpolation formula:

$$f(x) = \sum_{i=1}^m c_i \prod_{j \neq i} \left(\frac{x - \lambda_j}{\lambda_i - \lambda_j} \right) = \sum_i c_i e_i(x)$$

will have the properties that $e_i(\lambda_k) = \delta_{ik}$ and $f(\lambda_i) = c_i$. Combine it with the above proposition, we have,

Proposition 4.8.5: *We have* $e_i(A) = P_{\lambda_i}$. *Thus we have* $A = \lambda_1 e_1(A) + \cdots + \lambda_m e_m(A)$.

Proof. Since we have P_{λ_i}'s are projections, we have $P_{\lambda_i}^2 = P_{\lambda_i}$ and $P_{\lambda_i} P_{\lambda_j} = 0$ for all distinct i, j. We deduce that

$$A^2 = \lambda_1^2 P_{\lambda_1} + \cdots + \lambda_m^2 P_{\lambda_m}$$
$$\ldots\ldots$$
$$A^k = \lambda_1^k P_{\lambda_1} + \cdots + \lambda_m^k P_{\lambda_m}.$$

It follows that

$$e_i(A) = e_i(\lambda_i) P_{\lambda_i} = P_{\lambda_i}. \qquad \blacksquare$$

We may consider the relation of normality of matrices and the minimal polynomials of matrices. We have,

Proposition 4.8.6: *Let* A *be a normal matrix over the complex field* $\mathbb{C}$ *with distinct eigenvalues* $\lambda_1, \ldots, \lambda_m$. *Then the minimal polynomial of* A *is* $m_A(\lambda) = \prod_i (\lambda - \lambda_i)$.

Proof. It is easy to check that $m_A(A) = 0$ and no polynomial of smaller degree will do. $\qquad \blacksquare$

Remark: One can prove that over real field $\mathbb{R}$, the minimal polynomial of a normal matrix has no multiple factor. $\qquad \blacksquare$

Exercises

(1) Let A be a normal matrix. Then A is self-adjoint (hermitian), skew-hermitian or unitary iff all the roots of the characteristic polynomial are real, pure imaginary or with absolute value 1 respectively.

(2) If S, T are two normal operators which commute, show that $S + T$, ST are normal operators.

(3) Let $V = \mathbb{C}^3$, and a linear operator T defined by $T(z_1, z_2, z_3) = (2z_2 + iz_3, 2z_1 + z_3, -iz_1 + z_2)$. Is T normal? Is T self-adjoint (hermitian)? Is T unitary? Can T be unitarily diagonalized?

(4) Let T be a normal operator. Show that $nullspace(T) = nullspace(T^*)$.

(5) Find a real matrix A which cannot be triangularized by real orthogonal matrices.

(6) Let A be the following matrix,

$$A = \begin{bmatrix} 3 & 1+i \\ 1-i & 1 \end{bmatrix}.$$

Find by direct computation the eigenvalues of it and the associated eigenvectors.

Chapter 5

Bilinear Forms and Decompositions

5.1 Bilinear Forms. Sylvester's Law of Inertia

In this section the field F may not be the real or complex field.

In the definition of inner product over real numbers, we require it to be a two-variable function that is linear in both variables. We have,

Definition Let U, V be F-vector spaces. A map B from $V \times U \mapsto F$ is said to be *bilinear* iff

(1) $B(v + v', u) = B(v, u) + B(v', u)$.
(2) $B(v, u + u') = B(v, u) + B(v, u')$.
(3) $B(\lambda v, u) = B(v, \lambda u) = \lambda B(v, u)$. ∎

Remark: If $F = \mathbb{C}$, we may start with conjugate linear as in the definition of inner product over complex numbers $\mathbb{C}$, i.e., we replace the third condition by

(3)$'$ $B(\lambda v, u) = B(v, \bar{\lambda} u) = \lambda B(v, u)$

and obtain many interesting results. This will not be covered in this book. ∎

Definition A bilinear form $B : V \times V \mapsto F$ is said to be *symmetric* if $B(v, u) = B(u, v)$, $\forall v, u \in V$. A bilinear form $B : V \times V \mapsto F$ is said to be *alternating* if $B(v, v) = 0$, $\forall v \in V$. A bilinear form $B : V \times V \mapsto F$ is said to be *anti-symmetric* if $B(v, u) = -B(u, v)$, $\forall v, u \in V$. ∎

Definition A bilinear form $B(v, u)$ is said to be *non-degenerate* iff $B(v, u) = 0$, $\forall u \in U$, then $v = 0$, iff for any $0 \neq v \in V$, there must be some $u \in U$ such that $B(v, u) \neq 0$. ∎

Remark: In the finite dimensional case, take any basis $\{v_1, v_2, \ldots, v_n\}$ of V and $\{u_1, u_2, \ldots, u_m\}$ of U. Let B be a bilinear form and $a_{ij} = B(v_i, u_j)$. Then we may associate the matrix $A = (a_{ij})$ to B. Let $v = \sum_i x_i v_i$, $u = \sum_i y_j u_j$, then $B(v, u) = (x_1, x_2, \ldots, x_n) A (y_1, y_2, \ldots, y_m)^T$. We shall call the matrix A, the matrix associated with the bilinear form B (with respect to the basis $\{v_1, v_2, \ldots, v_n\}$ and $\{u_1, u_2, \ldots, u_m\}$). Let C be the matrix form of B with respect to another basis $\{v_1', v_2', \ldots, v_n'\}, \{u_1', u_2', \ldots, u_m'\}$. Then $C = P^T A Q$ where $P = [v_1', v_2', \ldots, v_n']$, and $Q = [u_1', u_2', \ldots, u_m']$ the matrices with $P = $ column vectors $v_1', v_2', \ldots, v_n'$ and $Q = $ column vectors $u_1', u_2', \ldots, u_m'$. It is easy to see that B is symmetric iff $V = U$ and $A = A^T$, and B is anti-symmetric iff $V = U$ and $A = -A^T$.

Furthermore if $V = U$, $\{v_1, v_2, \ldots, v_n\} = \{u_1, u_2, \ldots, u_n\}$ and $\{v_1', v_2', \ldots, v_n'\} = \{u_1', u_2', \ldots, u_m'\}$, then $P = Q$, we say A, C are *congruent*. ∎

We have the following proposition,

Proposition 5.1.1: *Given any bilinear form $B(v, u)$ on $V \times V$, then $C(v, u) = B(v, u) + B(u, v)$ is a symmetric bilinear form, and $D(v, u) = B(v, u) - B(u, v)$ is anti-symmetric. If the characteristic of the field F is not 2, then $B = (1/2)(C + D)$. In other words, any bilinear form on $V \times V$ is a sum of a symmetric one and an anti-symmetric one, if the characteristic of the field F is not 2.*

Proof. It is obvious. ∎

Example 5.1: Certainly a symmetric bilinear form may not define a real inner product. For instance, let us consider $\mathbb{R}^2$. Let us use the standard coordinate, and define $B((x_1, x_2)^T, (y_1, y_2)^T) = -x_1 y_1$. Then B is clearly a symmetric bilinear form, while $B((1, 0)^T, (1, 0)^T) = -1$, $B((0, 1)^T, (0, 1)^T) = 0$, and it is not even positive semi-definite. So it is not an inner product. ∎

Example 5.2: Let F be a field of characteristic 2. Let $V = F^2$. The bilinear form B defined by $B([a_1, a_2]^T, [b_1, b_2]^T) = a_1 b_2$ is neither symmetric nor anti-symmetric. The bilinear form D defined by $D([a_1, a_2]^T, [b_1, b_2]^T) = a_1 b_1$ is symmetric, while not alternating. ∎

We will have the following definition,

Definition (Congruence): Two matrices A, C are said to be *congruent* iff there is an invertible matrix P such that $P^T A P = C$. ∎

We have the following concept for a bilinear form,

Definition Let V be finite dimensional with a basis $\{e_1, e_2, \ldots, e_n\}$. Let T be a bilinear form on $V \times V$, and A be the matrix associated with T. Then the rank of A is called the *rank* of T. Note that the rank is independent of the basis $\{e_1, e_2, \ldots, e_n\}$ (with respect to different basis, the associated matrices are congruent). ∎

Definition Given a symmetric bilinear form $B(v, u)$ on $V \times V$, we define the *quadratic form* $q(v)$ associated with $B(v, u)$ as

$$q(v) = B(v, v).$$ ∎

Then we have the following identity,

Proposition 5.1.2: *Let V be finite dimensional and the characteristic of the ground field F is not 2. Then for any symmetric bilinear form $B(v, u)$ on $V \times V$ is determined by the quadratic form $q(v)$ associated with it as,*

$$B(v, u) = (1/4)(q(v + u) - q(v - u)).$$

Proof. Trivial. ∎

Proposition 5.1.3: *Let V be finite dimensional and the characteristic of the ground field F is not 2. Then for any symmetric bilinear form $B(v, u)$ on $V \times V$, there is a basis $\{e_1, e_2, \ldots, e_n\}$ such that the matrix represents $B(v, u)$ is diagonal.*

Proof. Let $q(v)$ be the associated quadratic form of $B(v, u)$. If $q(v) = 0$ for all $v \in V$. Then $B(v, u) = 0$, $\forall v, u \in V$. That means $B = 0$. Therefore the matrix that represents B is the zero matrix and our proposition is done.

We shall deal with the case that there is a vector v, such that $B(v, v) = q(v) \neq 0$. Let $U = \{\alpha v : \alpha \in F\}$ and $W = U^\perp = \{w : B(v, w) = 0\}$. We claim that: (1) $U \cap U^\perp = \{0\}$, (2) $U + U^\perp = V$ (i.e., we claim $V = U \oplus U^\perp$).

(1) Let $w \in U \cap U^\perp$. Then $w = cv$ and $B(v, w) = 0 = B(v, cv) = cB(v, v)$ which implies $c = 0$ and $w = 0$.

(2) Let $u \in V$ be any element. Then $u - (B(v, u)/B(v, v))v = w \in U^\perp$. Therefore $u = (B(v, u)/B(v, v))v + w$.

Let $\dim(V) = n$, $U = \{cv : c \in F\}$ and $W = U^\perp$. Then $\dim(W) = n - 1$ and the restriction of B to W is a symmetric bilinear form. By the induction on the dimension, there is a basis $v_2, \ldots, v_n$ which makes the matrix represent the restriction of B to W in the diagonal form. We have $v, v_2, \ldots, v_n$ is the selected basis. ∎

Proposition 5.1.4: *Let V be finite dimensional over the complex field $\mathbb{C}$. Let B be a symmetric bilinear form on $V \times V$. Then there exists a basis $\{e_1, e_2, \ldots, e_n\}$ (there is no inner product, hence there is no sense to say if they form an orthonormal basis), such that the associated matrix $A = (a_{ij})$ is diagonal (i.e., $a_{ij} = 0$ if $i \neq j$), and $B(e_i, e_i) = 1$ for and only for $1 \leq i \leq r$, $B(e_j, e_j) = 0$ for all $j > r$ where r is the rank of A (hence of B).*

Proof. If $B = 0$, we take $r = 0$ and $e_1, \ldots, e_n$ be any basis. If $n = 1, B \neq 0$, let v be a basis, and $B(v, v) = a \neq 0$. Let $e_1 = v/\sqrt{a}$. Then it is easy to see that $B(e_1, e_1) = 1$. Our proposition is obvious.

In general, let $B \neq 0$. Then there must be v such that $q(v) = B(v, v) \neq 0$. We may *normalize* v as previously to get e_1 such that $B(e_1, e_1) = 1$. Let $U = \{ae_1 : a \in \mathbb{C}\}$ and $W = \{w : w \in V, B(e_1, w) = 0\}$. We **claim** $V = U \oplus W$.

(Proof of the Claim). Let $v \in V$ be any vector. Let $B(e_1, v) = b$. Let $u = be_1$ and $w = v - u$. Clearly that $u \in U$, and $B(e_1, w) = B(e_1, v) - B(e_1, u) = b - bB(e_1, e_1) = 0$ and $w \in W$. Therefore $V \subset U + W$. On the other hand, let $v \in U \cap W$. Then $v = be_1$ since $v \in U$ and $0 = B(e_1, v) = B(e_1, be_1) = bB(e_1, e_1) = b$, in other words, $b = 0$ and $v = 0$. We proved the claim.

It is easy to see that the restriction of B to $W \times W$ is a symmetric bilinear form on W. By an induction on the dimension of the space, we know that there exists a basis $e_2, \ldots, e_n$ for W such that the statement of the proposition is satisfied. Certainly we have $B(e_1, e_i) = 0$ for all $i \geq 2$. Henceforth $e_1, e_2, \ldots, e_n$ is the basis we are looking for. ∎

Proposition 5.1.5 (Sylvester's[1] Law of Inertia): *Let V be finite dimensional over the real field $\mathbb{R}$, and B a symmetric bilinear form on $V \times V$. Then there exists a basis $\{e_1, e_2, \ldots, e_n\}$ (there is no inner product, hence there is no sense to say if they form an orthonormal basis), such that the associated matrix $A = (a_{ij})$ is diagonal (i.e., $a_{ij} = 0$ if $i \neq j$), $B(e_i, e_i) = 1$ for and only for $1 \leq i \leq p \leq r$, and $B(e_i, e_i) = -1$ for and only for $p + 1 \leq i \leq r$ where r is the rank of A (hence of B). Then the numbers r, p are independent of the basis. Furthermore, let $q(v)$ be the associated quadratic form, then $q(v)$ can be written as*

$$q(v) = \sum_{i=1}^{p} v_i^2 - \sum_{i=p+1}^{r} v_i^2$$

where with respect to some (fixed) basis of V, the coordinates of v is $(v_1, v_2, \ldots, v_n)^T$.

Proof. By imitating the proof of the preceding proposition, we can easily show that there is a basis $\{e_1, e_2, \ldots, e_n\}$ with the stated property. The only thing we have to prove is the number p which is independent of the basis. We know r is the rank which is independent of the basis.

---

[1]Sylvester, J. English mathematician. 1814–1897.

Let V^+ be the subspace spanned by $\{e_1, \ldots, e_p\}$, V^- be the subspace spanned by $\{e_{p+1}, \ldots, e_r\}$, and $V^\perp$ be the subspace spanned by $\{e_{r+1}, \ldots, e_n\}$. Then clearly we have $V = V^+ \oplus V^- \oplus V^\perp$. Note that B is positive-definite on V^+.

Let W be any subspace of V such that the restriction of B to W is positive-definite, we **claim** that any linearly dependent vectors $w \in W$, $v_- \in V^-$, and $v_\perp \in V^\perp$ must be all zeroes.

(Proof of the Claim). Let $w + v_- + v_\perp = 0$. Then we have, since $B(w, v_\perp) = 0$ for all w,

$$B(w, w + v_- + v_\perp) = B(w, w) + B(w, v_-) + B(w, v_\perp)$$
$$= B(w, w) + B(w, v_-) = 0$$
$$B(v_-, w + v_- + v_\perp) = B(v_-, w) + B(v_-, v_-) + B(v_-, v_\perp)$$
$$= B(v_-, v_-) + B(w, v_-) = 0.$$

We conclude that $B(w, w) = B(v_-, v_-)$. On the other hand, we have $B(w, w) \geq 0 \geq B(v_-, v_-)$. Therefore $B(w, w)^\bullet = B(v_-, v_-) = 0$ and $w = v_- = 0$. Now we conclude $v_\perp = 0$ and finish the proof of our claim.

From the elementary theory of linear algebra, we know $\dim(W) \leq \dim(V^+)$. Let $V = V_1^+ \oplus V_1^- \oplus V_1^\perp$ be another decomposition. Then we may use V_1^+ as W and conclude $\dim(V_1^+) \leq \dim(V^+)$. After we switch the roles of these two decompositions, we have $\dim(V_1^+) \geq \dim(V^+)$, and hence $\dim(V_1^+) = \dim(V^+) = p$. ∎

Corollary 5.1.6: *A symmetric bilinear form over $\mathbb{R}^n \times \mathbb{R}^n$ is non-degenerate iff $r = n$.* ∎

Definition Let the *signature* of $B - (p, (r \quad p)) - (p, q)$, where $q = r - p$. ∎

Corollary 5.1.7: *Two real symmetric $n \times n$ matrices A, B are congruent iff their signatures are identical.* ∎

In the case of a finite dimensional real vector space V and a symmetric bilinear form B on $V \times V$, the rank and the signature of B uniquely determine the numbers of $+1, -1$ on the diagonal for a diagonal form of the matrix,

Remark: Let A be a real symmetric matrix which can be *straightly reduced to a diagonal form* (the definition is in Section 4.6). In the case that A cannot be *straightly reduced to diagonal form*, note that $A - \epsilon I$, where ϵ has to avoid $n!$ many real eigenvalues of all principal minors of A, can be straightly reduced to diagonal form. We may replace A by $A - \epsilon I$ and apply the following arguments to $A - \epsilon I$, then we have two forms

$$L^T A L = D$$
$$O^{-1} A O = O^T A O = \Lambda.$$

According to Sylvester's theorem, the numbers of $+, -, 0$ are the same for both expressions. We may use the information to find all eigenvalues λ. Let us consider the following example,

$$A = \begin{bmatrix} 7 & 1 & 1 \\ 1 & 5 & 1 \\ 1 & 1 & 9 \end{bmatrix}.$$

(1) We verify that A is positive-definite, hence all eigenvalues λ_i are positive. (2) Let us consider $B = A - 10I$. We find that after *straightly reducing* B to its diagonal form, it has 3 negative pivots, -3, $-14/3$, $-22/21$, henceforth 3 negative eigenvalues $\lambda_1 - 10$, $\lambda_2 - 10$, $\lambda_3 - 10$. Therefore, all eigenvalues λ_i of A are between 0 and 10. (3) Let us consider $C = A - 5I$. Then we find that the pivots are 2, $-1/2$, 4. Therefore there are 2 eigenvalues λ_2, λ_3 of A which are bigger than 5, and λ_1 less than 5. (4) We may bisect the intervals and proceed like these, and find a good approximation of eigenvalues. (5) We may use a *Matlab* program to determine the eigenvalues are $\lambda = 4.5107, 6.7108, 9.7785$.

The above was the method to approximate the eigenvalues before 1960. Now we use the much faster QR method of Chapter 6. ∎

Proposition 5.1.8: *Let the characteristic of the ground field $\neq 2$. Then a bilinear form B is alternating iff it is anti-symmetric.*

Proof. In any characteristic we always have alternative $\Longrightarrow$ $B(v, v) = q(v) = 0 \Longrightarrow B(v, u) = 0 \Longrightarrow$ anti-symmetric. If the

characteristic is not 2, then $B(v, u) = -B(u, v) \implies B(v, u) = 0 \implies$
$B(v, v) = 0 \implies$ alternative. ∎

We have the following theorem about the form of an anti-symmetric bilinear form,

Proposition 5.1.9: *Let V be a finite dimensional vector space of dimension n over a field F with characteristic $\neq 2$, and B a non-degenerate anti-symmetric bilinear form on $V \times V$. Then n is even and $n = 2m$, and there exists a basis $\{e_1, e_2, \ldots, e_{2m}\}$ such that the associated matrix $A = (a_{ij})$ is as follows,*

$$\begin{bmatrix} 0 & 1 & 0 & 0 & 0 & 0 & \cdots & 0 \\ -1 & 0 & 0 & 0 & 0 & 0 & \cdots & 0 \\ 0 & 0 & 0 & 1 & 0 & 0 & \cdots & 0 \\ 0 & 0 & -1 & 0 & 0 & 0 & \cdots & 0 \\ \cdots & \cdots & \cdots & \cdots & \cdots & \cdots & \cdots & \cdots \\ \cdots & \cdots & \cdots & \cdots & \cdots & \cdots & \cdots & \cdots \\ 0 & 0 & 0 & 0 & 0 & 0 & 0 & 1 \\ 0 & 0 & 0 & 0 & 0 & 0 & -1 & 0 \end{bmatrix}$$

In other words, on the $(2i-1)$th row, all entries are zeroes, except at the $2i$th entry which is 1, and on the $2i$th row, all entries are zeroes except at the $(2i-1)$th entry which is -1.

Proof. Since B is non-degenerate, there must be a vector v such that for some u, $B(v, u) = a \neq 0$. Clearly $B((1/a)v, u) = 1$. Let $v_1 = (1/a)v$, and $v_2 = u$. Note that v_1, v_2 are linearly independent. Then we have $B(v_1, v_2) = 1, B(v_2, v_1) = -1$. Let U be the subspace generated by v_1, v_2 and $W = \{w : B(v_1, w) = 0, B(v_2, w) = 0\}$. We **claim** that $V = U \oplus W$. Given any $v \in V$. Let $b = B(v_1, v)$, $c = B(v_2, v)$. Then it is easy to see that $B(v - cv_1 + bv_2, v_i) = 0$ for $i = 1, 2$. Thus $v - cv_1 + bv_2 \in W$ and $V \subset U + W$. Let $v \in U \cap W$. Let $v = cv_1 - bv_2$. Then $B(v, v_i) = 0$. It is easy to show that $b = c = 0$. Our claim is proved.

If $W \neq 0$, then we **claim** that the restriction of B to W is anti-symmetric and non-degenerate. It is trivial to see that it is anti-symmetric. We shall show that it is non-degenerate. Let $0 \neq w \in W$.

Then $B(w, u) = 0$ for all $u \in U$. Therefore, there must be a $w' \in W$ such that $B(w, w') \neq 0$. Hence the restriction of B is non-degenerate.

Now we may apply the mathematical induction to W, and deduce that there is a basis $v_2, v_3, \ldots, v_{2n}$ for W, and $v_1, v_2, \ldots, v_{2n}$ a basis for V, such that the matrix form of B with respect to it is of the form in the proposition. ∎

Corollary 5.1.10: *Let V be finite dimensional over a field K of dim n with characteristic $\neq 2$, and B an anti-symmetric bilinear operator on $V \times V$. Then $n = 2m + r$, and there exists a basis $\{e_1, e_2, \ldots, e_{2m}, e_{2m+1}, \ldots, e_{2m+r}\}$ such that the associated matrix $A = (a_{ij})$ is as follows,*

$$
\begin{bmatrix}
0 & 1 & 0 & 0 & 0 & 0 & \ldots & 0 & 0 & 0 \\
-1 & 0 & 0 & 0 & 0 & 0 & \ldots & 0 & 0 & 0 \\
\ldots & \ldots & \ldots & \ldots & \ldots & \ldots & \ldots & \ldots & \ldots & \ldots \\
\ldots & \ldots & \ldots & \ldots & \ldots & \ldots & \ldots & \ldots & \ldots & \ldots \\
0 & 0 & 0 & 0 & 0 & 0 & 0 & 1 & 0 & 0 \\
0 & 0 & 0 & 0 & 0 & 0 & -1 & 0 & 0 & 0 \\
0 & 0 & \ldots & 0 & 0 & 0 & 0 & 0 & 0 & 0 \\
0 & 0 & 0 & \ldots & 0 & 0 & 0 & 0 & 0 & 0
\end{bmatrix}.
$$

In other words, for $i \leq m$ on the $(2i-1)$th row, all entries are zeroes, except at the $2i$th entry which is 1, and on the $2i$th row, all entries are zeroes except at the $(2i-1)$th entry which is -1 and for $i > 2m$, all entries on the ith row are zeroes.

Proof. It is trivial. ∎

Corollary 5.1.11: *The rank of an anti-symmetric matrix is even and its determinant is a square.*

Proof. Trivial. ∎

Example 5.3: The anti-symmetric bilinear form is significant in modern mathematics and the sciences. In the 2-dimensional real vector space $\mathbb{R}^2$, we have the area $A(v_1, v_2) = \det[v_1, v_2]$ as a well-known anti-symmetric bilinear form. In fact, in any $2n$-dimensional real vector space, for any two real vectors v, u, we may use the area of the parallelogram spanned by the two vectors to define an anti-symmetric bilinear form. We may discuss a differential geometry base on a symmetric bilinear differential form ω on the tangent space. And we may discuss a *symplectic geometry* based on an anti-symmetric bilinear differential form w on the tangent space. In the case of *symmetric bilinear form*, the important concept is the *length* (which is allowed to be imaginary). In the case of *symplectic geometry* the important concept is the *area*.　　■

We have the following examples,

Example 5.4: Let $\{a_{ij}\}$ be a set of symbols. Let A be an anti-symmetric matrix over $S[\{a_{ij}\}]$ as follows, and their determinants are squares

$$\det(A) = \det \begin{bmatrix} 0 & a_{12} \\ -a_{12} & 0 \end{bmatrix} = a_{12}^2$$

$$\det(A) = \det \begin{bmatrix} 0 & a_{12} & a_{13} & a_{14} \\ -a_{12} & 0 & a_{23} & a_{24} \\ -a_{13} & -a_{23} & 0 & a_{34} \\ -a_{14} & -a_{24} & -a_{34} & 0 \end{bmatrix}$$

$$= (a_{12}a_{34} - a_{13}a_{24} + a_{14}a_{23})^2.　　■$$

From the above examples, one sees that if A is anti-symmetric, it is plausible that $\det(A)$ is a square. In fact, we have the following proposition,

Proposition 5.1.12 (Muir[2], 1882): *Let the characteristic of the ground field not be 2. Let a_{ij} be symbols for all $2n > i > j > 0$, and $a_{ji} = -a_{ij}$. Let $A = (a_{ij})$. Then the anti-symmetric matrix is with determinant that is a square, i.e.,*

$$\det(A) = Pf(A)^2.$$

The above function $Pf(A)$ is determined by the above equation with the extra requirement that the coefficient of $a_{1,2}, a_{3,4}, \ldots, a_{2n-1,2n}$ is 1 (otherwise it might be -1).

---

[2]Sir Thomas Muir. Scottish mathematician. 1844–1934.

Proof. It suffices to show that over the field $Q(\{a_{ij}\})$ the above formula is true. Since $A = P^T D P$ where D is of the form in the preceding proposition and $\det(D) = 1$ or 0. Therefore, we have $\det(A) = \det(P)^2$ or 0. We claim that $\det(P)$ is a polynomial in $\{a_{ij}\}$ with integer coefficients, i.e., $\det(P) \in \mathbb{Z}[\{a_{ij}\}]$.

If $\det(D) = 0$, there is nothing to be proved. Assume that $\det(D) = 1$. Let $\det(P) = f/g$ with $f, g \in \mathbb{Z}[\{a_{ij}\}]$ and co-prime. Then we have $g \det(P) = f$. Due to the UFD property of $\mathbb{Z}[\{a_{ij}\}]$, we conclude that $g = \pm 1$. ∎

Definition 5.13: For any anti-symmetric matrix $A = (b_{ij})$, we define $Pf(A)$ to be the specialization of the above Pfaffian[3] with $a_{ij} \mapsto b_{ij}$. The polynomial $Pf(A)$ (the *Pfaffian of A*) (in 1852, Cayley named it after Pfaff) for any anti-symmetric matrix A is defined. ∎

Exercises

(1) Let $V = \mathbb{R}^2$, and the quadratic form $q = x_1^2 + x_1 x_2 + x_2^2$. Find the corresponding bilinear form B.

(2) Let $V = \mathbb{R}^4$. Find a symmetric bilinear form with signature $(3,1)$ and a symmetric bilinear form with signature $(2,2)$.

(3) Show that if the signature of a symmetric bilinear form B on an n-dimensional vector space V over $\mathbb{R}$ the real numbers are $\pm n$, then the form B is either positive-definite, i.e., $B(v,v) > 0$ for all $v \neq 0$ or $-B$ is positive-definite.

(4) Show that if the signature of a symmetric bilinear form B on an n-dimensional vector space V over $\mathbb{R}$ the real numbers are not $\pm n$, then there must be a vector $v \neq 0$, such that $B(v,v) = 0$.

(5) Show that any symmetric bilinear form over $\mathbb{R}^4$ with signature $(4,0)$ is positive-definite.

(6) Show that if V is of odd dimension over a field K with characteristic $\neq 2$, then any anti-symmetric bilinear form B is with determinant 0.

---

[3]Pfaff, J. German mathematician. 1765–1825.

(7) Given a bilinear operator A_1 defined over a vector space $V_1 \times V_1$ over a field K with characteristic $\neq 2$ and a bilinear operator A_2 defined over a vector space $V_2 \times V_2$ over the same field K. We define $A_1 \oplus A_2$ as the bilinear operator defined over $(V_1 \oplus V_2) \times (V_1 \oplus V_2)$ as $(A_1 \oplus A_2)((v_1, v_2), (u_1, u_2)) = A_1(v_1, u_1) + A_2(v_2, u_2)$. Show that if both A_1 and A_2 are anti-symmetric, then $A_1 \oplus A_2$ is anti-symmetric, and $Pf(A_1 \oplus A_2) = Pf(A_1)Pf(A_2)$.

(8) Let $C = BAB^T$. Show that $Pf(C) = \det(B)Pf(A)$.

(9) Let $A_{\hat{i}\hat{j}}$ be the matrix obtained from A by deleting the ith row and column and jth row and column. Show that $Pf(A) = \sum (-1)^j a_{1j} Pf(A_{\hat{1}\hat{j}})$.

5.2   Groups Preserving Bilinear Forms. Special Relativity

One way to study the natural sciences is the way of Emmy Noether who deduced all physical conservative laws in high energy physics (physics on elementary particles), such as the conservative law of energy, momentum, angular momentum etc., as invariants from the actions of some groups. Or we may start with the physical laws first, and then try to find the group actions which preserve those quantities.

We may consider a bilinear form as the fundamental defining quantity, and consider all linear operators which preserve them. We have the following definition,

Definition Let $f(v, u)$ be a bilinear form on $V \times V$, and T a linear operator from V to V. If $f(Tv, Tu) = f(v, u)$, for all v, u, then we say T preserves the bilinear form $f(v, u)$.   ∎

We have the following proposition,

Proposition 5.2.1: *Let V be a finite dimensional vector space. Let B be a non-degenerate bilinear form on $V \times V$. Then all linear operators T which preserve B, i.e., $B(Tv, Tu) = B(v, u)$ for all $v, u \in V$, form a group.*

Proof. We have to show that every such linear operator T has an inverse. In the finite dimensional case, we have to show that the nullspace of T is $\{0\}$.

Let $v \in$ the nullspace, i.e., $T(v) = 0$. Then $0 = B(Tv, Tu) = B(v, u)$ for all u. Hence $v = 0$. Our claim is proved.   ∎

According to Proposition 5.1.5, over the real field $\mathbb{R}$ we may assume a symmetric bilinear form on a finite dimensional $V \times V$ is diagonal with diagonals ± 1, 0.

We shall define,

Definition Let V be an n-dimensional real space. Let B be a non-degenerate symmetric bilinear form on $V \times V$. Then the group of linear operators T which preserves B, i.e., $B(Tv, Tu) = B(v, u)$ for all $v, u \in V$, is called the *indefinite orthogonal group* $O_{(p,q)}$ where $p, q \ (= n - p)$ are the signatures of B. Note that a real symmetric bilinear form B determines and is uniquely determined by a quadratic form $q(v)$. We shall call the indefinite orthogonal group $O_{(p,q)}$ the *group of $q(v)$*.   ∎

We have the following definition,

Definition Let V be a K-vector space of dimension n. The group which preserves a non-degenerate anti-symmetric bilinear form B defined over $V \times V$ is called a *symplectic group* $Sp(2n, K)$.   ∎

It goes without saying that when $n = p$ or $q = 0$, then $O_{(p,q)} = O_n$. In the case of $V = \mathbb{R}^n$, every element in O_n is determined by an orthonormal basis $\{v_1, v_2, \ldots, v_n\}$ of $\mathbb{R}^n$; we may pick v_1 as any vector on the unit $(n-1)$-sphere, which is $(n-1)$-dimensional. Then we pick up the plane which is perpendicular to v_1, and in it we find an $(n-2)$-dimensional sphere, and we pick up any vector v_2 on the $(n-2)$-sphere. Inductively we may continue to compute the dimension of O_n which is summed as follows,

$$\sum_{i=1}^{n-1} i = n(n-1)/2.$$

So O_n is a *Lie group* (a group which is a differentiable manifold with group operation continuous with respect to the smooth structure) of dimension $n(n-1)/2$. Similar arguments will show that $O_{(p,q)}$ is a Lie group of dimension $n(n-1)/2$ where $n = p + q$.

We have the following examples,

Example 5.5 (Orthogonal Group): Let the field be real numbers $\mathbb{R}$, $V = \mathbb{R}^n$. If the symmetric bilinear form on $V \times V$ is with $p = n$, $q = 0$, then the bilinear form is positive-definite and define an inner product. Therefore the group O_n is the usual orthogonal group which fixes *length* (hence angles). ∎

Example 5.6 (Galileo[4] Group): If we start with classical physics (Newtonian[5] physics), then we have formula of motion

$$x^* = x - vt.$$

The Galileo group consists of transformations (Galileo group) of the form,

$$x^* = x - vt$$
$$t^* = t$$

will keep the law of motion to be invariant. From there, we get that the separation of space and time, and a flat space, a steady flow of time. Those were the two categories of Aristotle and were called *forms of intuition* by a famous philosopher, Kant.[6] ∎

Towards the end of the 19th century, we have the following *Maxwell's[7] equations* for electromagnetic fields (here we use the distance which the light travels in one second as the unit of length 1),

$$\nabla \cdot E = \rho\epsilon_0, \quad \text{Gauss's law for electricity}$$

$$\nabla \cdot B = 0, \quad \text{Gauss's law for magnetism}$$

$$\nabla \times E = -\frac{\partial B}{\partial t}, \quad \text{Faraday equation}$$

$$\nabla \times B = \mu_0 J + \mu_0\epsilon_0\frac{\partial E}{\partial t} \quad \text{Ampere's circuital law}$$

---

[4]Galileo, G. Italian polymath. A central figure of modern science. 1564–1642.

[5]Newton, I. English mathematician, astronomer, theologian, author and physicist. 1643–1727.

[6]Kant, I. German philosopher. 1724–1804.

[7]Maxwell, J. Scottish scientist. Founder of electromagnetic theory. 1831–1879.

where Faraday[8] and Ampere[9] are famous physicists. We use the common symbols as listed below, and

> ρ is the charge density
> E is the electric field
> B is the magnetic induction
> ϵ_0 is the electric constant
> μ_0 is the magnetic constant
> J is the electric current.

Example 5.7 (Electromagnetic Wave): In the vacuum we have $\rho = J = 0$, and $\mu_0\epsilon_0 = c^2$ as the square of the speed of light, which is the $c^2 = 1$ in our convention. Then the Maxwell's equations are reduced to

$$\nabla \cdot E = 0$$
$$\nabla \cdot B = 0$$
$$\nabla \times E = -\frac{\partial B}{\partial t}$$
$$\nabla \times B = \frac{\partial E}{\partial t}.$$

Using the relations $\nabla \times (\nabla \times \bullet) = (\nabla \cdot \nabla - \nabla^2)\bullet$ and the commutative of $\frac{\partial}{\partial t}$ and $\nabla\cdot$, we have the following wave equations,

$$\nabla^2 E = \frac{\partial^2 E}{(\partial t)^2}$$
$$\nabla^2 B = \frac{\partial^2 B}{(\partial t)^2}.$$

Here the speed of the electromagnetic wave is $1 =$ the speed of light. ∎

---

[8]Faraday, M. English scientist. 1791–1867.
[9]Ampere, A. French mathematician and scientist. 1775–1836.

Lorentz[10] found that the above equations are invariant under the following substitutes, L_v as the *Lorentz boost*,

$$x^* = \frac{x - vt}{\sqrt{1 - v^2}} = \frac{x - vt}{\Delta}$$

$$y^* = y$$

$$z^* = z$$

$$t^* = \frac{t - vx}{\sqrt{1 - v^2}} = \frac{t - vx}{\Delta}.$$

Note that the group is different from the Galileo group. This shows that there is a deep split between classic mechanics and electromagnetic theory. In other words, their groups and hence their geometries are different. Later on, Einstein[11] postulated that in *special relativity*, the speed of light is invariant under all uniformly moving frames and unified the classical mechanics and electromagnetic theory using the Lorentz transformations. Minkowski[12] discussed Einstein's special relativity from a mathematical point of view and showed that it could be reinterpreted as a group which preserves the following quadratic form,

$$q = t^2 - x^2 - y^2 - z^2.$$

It follows that, we have $O_{1,3}$ as part of the isometric group (i.e., Poincaré group) of the above quadratic form. Certainly, it is interesting to find all linear operators which preserves the above quadratic form. Let σ be an element which is in the linear isometric group which preserves the above quadratic form. Let $x' = \sigma(x)$, $y' = \sigma(y)$, $z' = \sigma(z)$, $t' = \sigma(t)$, then we may select y', z' independent of the variable t, a further linear transformation τ will send y' to y and z' to z. We may assume that the linear transform is

---

[10]Lorentz, H. Dutch physicist. 1902 Nobel Prize winner. 1853–1928.
[11]Einstein, A. German-born, citizen of several countries. He made fundamental contributions to *special relativity*, *general relativity*, *quantum theory* and *Brownian motions*. In 1921, he received the Nobel Prize in Physics. 1879–1955.
[12]Minkowski, H. German mathematician. 1854–1909.

of the form,

$$x^* = f_1(x, y, z, t)$$
$$y^* = y$$
$$z^* = z$$
$$t^* = f_4(x, y, z, t).$$

Proposition 5.2.2: *Let us consider the following linear substitution σ (named as a Lorentz transformation) which preserves the above quadratic form $q = t^2 - x^2 - y^2 - z^2$,*

$$x^* = f_1(x, y, z, t)$$
$$y^* = y$$
$$z^* = z$$
$$t^* = f_4(x, y, z, t).$$

Then it is a **Lorentz boost**.

Proof. Let us rewrite σ as,

$$x^* = g_1(x, t) + h_1(y, z)$$
$$y^* = y$$
$$z^* = z$$
$$t^* = g_4(x, t) + h_4(y, z).$$

We shall show that a Lorentz transformation is a Lorentz boost.

Using the fact that L preserves the quadratic form q, one proves easily that we have

$$x^2 - t^2 = g_1(x, t)^2 + 2g_1(x, t)h_1(y, z) + h_1(y, z)^2 - g_4(x, t)^2$$
$$- 2g_4(x, t)h_4(y, z) - h_4(y, z)^2.$$

Computing the terms of y, z, we can easily conclude that $h_1 = \pm h_4$. If $h_1 \neq 0$, then we must have $g_1 = \pm g_4$, then x^*, y^*, z^*, t^* are linearly dependent which is impossible. Therefore we conclude that $h_1 = h_4 = 0$. Disregarding the y, z-coordinates, we may study further the following equations,

$$x^* = ax + bt$$
$$t^* = cx + dt$$

which preserve the following quadratic form

$$q = t^2 - x^2.$$

(Mathematical deduction) We have

$$t^2 - x^2 = (t - x)(t + x) = (t^* - x^*)(t^* + x^*).$$

It is easy to deduce that

$$t^* - x^* = \lambda(t - x)$$
$$t^* + x^* = \lambda^{-1}(t + x)$$

where $\lambda > 0$, since the linear operator will not change the directions of t and x. We may write $\lambda = e^\phi$. Remember that

$$\sinh(\phi) = \frac{e^\phi - e^{-\phi}}{2}$$

$$\cosh(\phi) = \frac{e^\phi + e^{-\phi}}{2}.$$

Then the linear operator can be written as

$$x^* = \cosh(\phi)x - \sinh(\phi)t$$
$$t^* = -\sinh(\phi)x + \cosh(\phi)t.$$

Or we may express them in the matrix form,

$$\begin{bmatrix} x^* \\ t^* \end{bmatrix} = \begin{bmatrix} \cosh(\phi) & -\sinh(\phi) \\ -\sinh(\phi) & \cosh(\phi) \end{bmatrix} \begin{bmatrix} x \\ t \end{bmatrix}.$$

Define Δ, v by the following formula,

$$\cosh(\phi) = \frac{1}{\Delta} = \frac{e^\phi + e^\phi}{2}$$

$$\sinh(\phi) = \frac{v}{\Delta} = \frac{e^\phi - e^\phi}{2}.$$

We have

$$v = \tanh(\phi) = \frac{\sinh(\phi)}{\cosh(\phi)}.$$

It is easy to see that

$$\cosh^2(\phi) - \sinh^2(\phi) = 1 = \frac{1}{\Delta^2} - \frac{v^2}{\Delta^2}$$

or

$$\left[\begin{array}{l} \Delta = \sqrt{1 - v^2} \\ x^* = \dfrac{x - vt}{\Delta} \\ t^* = \dfrac{t - vx}{\Delta} \end{array}\right].$$

We have a *Lorentz boost*, which is a *rotation* in the x-t spacetime. ∎

Example 5.8 (Wave Equation in Differential Forms with Special Relativity):

In *differential geometry*, we will formulate Maxwell's equations for electromagnetic wave (cf. Example 5.7) as follows,

$$\frac{\partial \mathbf{B}}{\partial t} = -\nabla \times \mathbf{E}, \quad \nabla \cdot \mathbf{B} = 0 \tag{1}$$

$$\frac{\partial \mathbf{E}}{\partial t} = \nabla \times \mathbf{B}, \quad \nabla \cdot \mathbf{E} = 0 \tag{2}$$

where

$$\mathbf{B} = B_x \mathbf{i} + B_y \mathbf{j} + B_z \mathbf{k}$$
$$\mathbf{E} = E_x \mathbf{i} + E_y \mathbf{j} + E_z \mathbf{k}.$$

We define the *Faraday 2-form* $\mathbb{F}$ as

$$\mathbb{F} = \mathbb{E} + \mathbb{B}$$

where

$$\mathbb{E} = E_x dt \wedge dx + E_y dt \wedge dy + E_z dt \wedge dz$$
$$\mathbb{B} = B_x dz \wedge dy + B_y dx \wedge dz + B_z dy \wedge dx.$$

Hodge Star

Given an n-dimensional manifold, it is a map which establishes a duality between k-form and $(n - k)$-form. Let $\sigma = (i_1, i_2, \ldots, i_n)$, then the Hodge star $*$ is defined as

$$*(dx_{i_1} \wedge \cdots \wedge dx_{i_k}) = \text{sign}(\sigma)\epsilon_{i_1}\epsilon_{i_2}\cdots\epsilon_{i_k} \wedge dx_{i_{k+1}} \cdots \wedge dx_{i_n}.$$

Let us consider the Minkowski spacetime with signature $+ - --$. Then we have $\epsilon_t = -\epsilon_x = -\epsilon_y = -\epsilon_z = 1$, and

$$*(dt \wedge dx) = dz \wedge dy, \qquad *(dz \wedge dy) = -dt \wedge dx, \qquad (3)$$

$$*(dt \wedge dy) = dx \wedge dz, \qquad *(dx \wedge dz) = -dt \wedge dy, \qquad (4)$$

$$*(dt \wedge dz) = dy \wedge dx, \qquad *(dy \wedge dx) = -dt \wedge dz. \qquad (5)$$

The 2-form $*\mathbb{F}$ is called *Maxwell* 2-form. It is easy to see that $*\mathbb{E} = \mathbb{B}$ and $*\mathbb{B} = -\mathbb{E}$.

Differential Forms for Maxwell's Equations

We will show that Maxwell's equations in vacuum are equivalent to the following equation,

$$d\mathbb{F} = 0, \quad d*\mathbb{F} = 0.$$

We shall only prove that $d\mathbb{F} = 0$ is equivalent to the two equations of (1) in this section. The proof that $d*\mathbb{F} = 0$ is equivalent to the two equations of (2) in this section is left to the reader. We have

$$d\mathbb{E} = dt \wedge \left\{ \left(\frac{\partial E_z}{\partial y} - \frac{\partial E_y}{\partial z} \right) dz \wedge dy + \left(\frac{\partial E_x}{\partial z} - \frac{\partial E_z}{\partial x} \right) dx \wedge dz \right.$$

$$\left. + \left(\frac{\partial E_y}{\partial x} - \frac{\partial E_x}{\partial y} \right) dy \wedge dx \right\}$$

$$d\mathbb{B} = dt \wedge \left(\frac{\partial B_x}{\partial t} dz \wedge dy + \frac{\partial B_y}{\partial t} dx \wedge dz + \frac{\partial B_z}{\partial t} dy \wedge dx \right)$$

$$+ \left(\frac{\partial B_x}{\partial x} + \frac{\partial B_y}{\partial y} + \frac{\partial B_z}{\partial z} \right) dx \wedge dy \wedge dz.$$

Therefore, we have $d\mathbb{F} = d\mathbb{E} + d\mathbb{B} = 0 \Leftrightarrow \nabla \times \mathbf{E} = -\frac{\partial \mathbf{B}}{\partial t}$ and $\nabla \cdot \mathbf{B} = 0$. Similarly the reader may prove that $d*\mathbb{F} = 0 \Leftrightarrow d\mathbb{B} - d\mathbb{E} = 0 \Leftrightarrow \nabla \times \mathbf{B} = \frac{\partial \mathbf{E}}{\partial t}$ and $\nabla \cdot \mathbf{E} = 0$.

Remark: In general, vacuum or not, we have $d\mathbb{F} = 0$ and $d * \mathbb{F} = *J$ where J is a 4-dimensional 1-form, with 3-space components corresponding to the ordinary current, and the time component the density of charge. The first equation is called the *homogeneous* one and the second one is the *inhomogeneous*. One may believe that this system of equations is not elegant enough. One may conjecture as Dirac does that one should change the equations to $d\mathbb{F} = M$, $d * \mathbb{F} = *J$, where M is similar to J in representing the *hypothetical Dirac magnetic current* with the time component being the density of the *hypothetical* Dirac monopoles of magnetics. ∎

Example 5.9 ($O_{p,n-p}$): In the previous example, we have studied $O_{(1,3)}$ which includes all of the orthonormal transformations in 3-space. Its dimension is 3 and all Lorentz boosts for one space dimension and one time dimension which is of dimension 3 again. It is easy to see that the total dimension is 6.

In general, let us consider $O_{(p,n-p)}$. There are orthogonal transformations on the first p-axis and last $(n-p)$-axis. The total dimensions for those parts are

$$\dim(O_p) = p(p-1)/2, \quad \dim(O_q) = q(q-1)/2.$$

Furthermore there are *Lorentz boost* type linear operators involving one of the first p-axis and one of the last q-axis. We have pq more dimensions. Therefore we have totally,

$$p(p-1)/2 + q(q-1)/2 + pq = (p+q)(p+q-1)/2$$
$$= n(n-1)/2. \quad ∎$$

Exercises

(1) Show that the speed of light cannot be exceeded, let the object E represent a standstill object, the object S represent an object moving with a speed $v < c$ the speed of light. We may further assume that $c = 1$ by changing the unit of length or time along the x-axis; which is associated with the Lorentz boost L_v, and P a third object moving relative to S with a speed $u < c = 1$, which is associated with a Lorentz boost L_u along the same x-axis.

Then the relative speed of the third object P relative to E is $L_u \circ L_v = L_w$. Show that $-1 < w < 1$.

(2) Let us use the notations of the preceding problem. Let $v = u = 2/3$. Find w.

(3) Prove Eqs. (3), (4) and (5).

(4) Show that $d * \mathbb{F} = 0$ iff Eq. (2) are satisfied for Maxwell's equations in vacuum.

(5) Let the quadratic form $q(t, x) = t^2 - x^2$ be given over R^2. Find its indefinite orthogonal group.

(6) Let the quadratic form $q(t, x) = t^2 - x^2 - y^2$ be given over R^3. Find its indefinite orthogonal group.

(7) Let the quadratic form $q(t, x) = t^2 - x^2 - y^2 - z^2$ be given over R^4. Find its indefinite orthogonal group.

5.3 Singular Value Decompositions. Its Applications

Since the time of Babylonia, we solve a quadratic equation by the method of *completing the square*, i.e., we rewrite an equation

$$f(x) = x^2 + bx + c = 0$$

as

$$(x - (b/2))^2 = b^2/4 - c = (1/4)(b^2 - 4c)$$

or

$$y^2 = \alpha.$$

If $b^2 - 4c$ is non-negative, then we take the square root of both sides, and find two roots of the original equation. On the other hand, if it is negative, then there were no real roots. The term $b^2 - 4c$ is called the *discriminant of the original equation*. It shows that if we express the equation in the form of $y^2 = \alpha$, then the number of entries may be reduced and the certain properties become obvious. For matrices, we are trying to do similar things.

Let A be any $m \times n$ matrix, then there exist invertible matrices C, D such that CAD which is the *normal form* of A

(cf. Theorem 1.1.11) is as,

$$CAD = \begin{bmatrix} 1 & 0 & \cdots & \cdots & 0 \\ \cdot & \cdot & \cdots & \cdots & 0 \\ \cdot & \cdot & 1 & \cdots & \cdots \\ \cdots & \cdots & \cdots & \cdots & \cdots \\ 0 & 0 & \cdots & \cdots & 0 \end{bmatrix}.$$

This form of A says many things about A.

Later we have the Smith Normal Form (cf. Section 2.4). Given $A = (a_{ij}) \in M_{n \times m}(R)$ where R is a P.I.D. and $A \neq (0)$. Then we may use the invertible maps to change A into a matrix $A' = (a'_{ij})$, i.e., there are invertible matrices C, D with $CAD = A' = (a'_{ij})$ such that

(1) $a'_{ij} = 0$ if $i \neq j$.

(2) $a'_{11}|a'_{22}|\cdots|a'_{\ell\ell}, \quad a'_{jj} \neq 0, \quad \forall j \leq \ell, \quad a'_{jj} = 0, \quad \forall j > \ell.$

Once we have the Smith Normal Form, we deduce many interesting results.

In the 19th century, the differential geometers created Riemann geometry by defining a Riemann metric ds as

$$ds^2 = \sum_{ij} a_{ij} dx_i dx_j, \qquad a_{ij} = a_{ji},$$

then classifying a quadratic form as,

$$\sum_{i=1,j=1}^{i=n,j=n} a_{ij} x_i x_j = x^T A x.$$

Note that A is a symmetric matrix. We may diagonalize it under the selections of orthonormal basis for $\{x_i\}$. Then we may read out many properties of the metric from the diagonal form.

Many differential geometers want to classify bilinear forms B on $V \times U$ where V, U are two finite dimensional vector spaces.

In differential geometry, the orthonormal basis are outstanding. We were allowed to alter basis using orthogonal matrices only. Beltrami[13] invented the singular value decomposition (see below) in 1873, and Jördan[14] independently discovered singular value decomposition (see below) in 1874. There are several other mathematicians with independent results.

We present *singular value decomposition* in general situations. In many situations, the matrix may not be the matrix of a linear operator, the domain and the range may be unrelated. Even worse, the matrix may not even be a linear transformation. It may be a rectangular expression of numerals of pictures or the *latent semantic index* of an article. We still want to put it in some standard decomposition. Let us look at the following example,

Example 5.10: Let us consider an example in *mechanics*. We want to analyze a motion represented by a *linear transformation* matrix A from $\mathbb{R}^2$ to $\mathbb{R}^2$ as follows,

$$A = \begin{pmatrix} 1 & 0.5 \\ 0 & 1 \end{pmatrix}.$$

The matrix A which does not change the x-axis while moving the second coordinate by a shearing force. We rewrite A as

$$A = O_1 \Sigma O_2^T$$

where O_i are orthogonal transformations and Σ is a diagonal matrix of positive numbers with $\lambda_1 = 9/8 + \sqrt{17}/8$, $\lambda_2 = 9/8 - \sqrt{17}/8$ on the diagonal. The above decomposition means that we may apply the matrix O_2^T which is a rotation + reflection, and then Σ which is a scaling change on the two axes, and then O_1 which is a rotation + reflection. Then end result will be a transformation of a unit circle to itself by Q_2^T and then we change by Σ the circle to an ellipse, and then rotate or reflect the ellipse to a similar ellipse. We have a better

---

[13]Beltrami, E. Italian mathematician. 1835–1900.
[14]Jördan, C. French mathematician. 1838–1922.

understanding of the motion represented by matrix A. The material which supports the motion should stand a stretch represented by λ_1 and a press represented by λ_2. ∎

In general, we may consider the *linear transformation* matrix A from $\mathbb{C}^n$ to $\mathbb{C}^m$ which may not be a square matrix. Let us use the standard inner product of $\mathbb{C}^n$, $\mathbb{C}^m$. We may select suitable orthonormal bases for $\mathbb{C}^n$, $\mathbb{C}^m$. In other words we may modify the matrix A from the left by U^* and from the right by V and consider U^*AV where U is an $m \times m$ unitary matrix and V is an $n \times n$ unitary matrix. It turns out that we may make U^*AV *diagonal* with positive real numbers on the diagonal (see below for a proof). If it is true, then we have

$U^*AV = \Sigma$, diagonal (for a definition see below) or $A = U\Sigma V^*$.

Or equivalently, with u_i column vectors of U and v_i column vectors of V, we have

$$Av_i = \delta_i u_i, \quad \text{for } i = 1, \ldots, \min(m, n), \text{ and } \delta_i \geq 0.$$

We define,

Definition An $m \times n$ matrix $A = (a_{ij})$ is said to be *diagonal* iff $a_{ij} = 0$ for all $i \neq j$. ∎

Example 5.11: The following matrices are both diagonal,

$$A = \begin{bmatrix} 3 & 0 & 0 \\ 0 & 1 & 0 \end{bmatrix}, \quad B = \begin{bmatrix} 3 & 0 \\ 0 & 1 \\ 0 & 0 \end{bmatrix}.$$

∎

Definition Given an $m \times n$ matrix A defining a map from $\mathbb{C}^n$ to $\mathbb{C}^m$. A *singular value decomposition* of A is written as $A = U\Sigma V^*$ where U, V are unitary matrices, and Σ is diagonal with non-negative diagonals. In other words, we have an orthonormal basis $v_1, \ldots, v_n$ for $\mathbb{C}^n$ and an orthonormal basis $u_1, \ldots, u_m$ for $\mathbb{C}^m$. Then

$$Av_i = \delta_i u_i, \quad \text{for } i = 1, \ldots, \min(m, n), \text{ and } \delta_i \geq 0.$$

∎

We have the following computational example,

Example 5.12: Let us consider the following 4×5 matrix A,

$$A = \begin{bmatrix} 1 & 0 & 0 & 0 & 2 \\ 0 & 0 & 3 & 0 & 0 \\ 0 & 0 & 0 & 0 & 0 \\ 0 & 4 & 0 & 0 & 0 \end{bmatrix}.$$

A singular value decomposition of A is given by

$$U = \begin{bmatrix} 0 & 0 & 1 & 0 \\ 0 & 1 & 0 & 0 \\ 0 & 0 & 0 & 1 \\ 1 & 0 & 0 & 0 \end{bmatrix}, \quad \Sigma = \begin{bmatrix} 4 & 0 & 0 & 0 & 0 \\ 0 & 3 & 0 & 0 & 0 \\ 0 & 0 & \sqrt{5} & 0 & 0 \\ 0 & 0 & 0 & 0 & 0 \end{bmatrix},$$

$$V^* = \begin{bmatrix} 0 & 1 & 0 & 0 & 0 \\ 0 & 0 & 1 & 0 & 0 \\ \sqrt{0.2} & 0 & 0 & 0 & \sqrt{0.8} \\ 0 & 0 & 0 & 1 & 0 \\ \sqrt{0.8} & 0 & 0 & 0 & -\sqrt{0.2} \end{bmatrix}. \qquad \blacksquare$$

Theorem 5.3.1: *Let A be an $m \times n$ matrix from W_1 to W_2 over either the real field $\mathbb{R}$ or the complex field $\mathbb{C}$. Then there exists a decomposition of the form,*

$$A = U\Sigma V^*$$

where U is an $m \times m$ unitary matrix and V is an $n \times n$ unitary matrix, and Σ is an $m \times n$ diagonal matrix with non-negative real numbers on the diagonal.

Proof. We know that for any matrix A, A^*A and AA^* satisfy

$$(AA^*)^* = AA^* \quad (A^*A)^* = A^*A,$$

hence they are self-adjoint, furthermore

$$x^*(AA^*)x = (A^*x)^*(A^*x) \quad x^*(A^*A)x = (Ax)^*(Ax),$$

therefore they are positive semi-definite. Let us assume $m \geq n$ and consider A^*A. According to the spectral theorem of self-adjoint matrix, there is an $n \times n$ unitary matrix V such that V^*A^*AV is diagonal.

Now we will give two different proofs of the theorem,

(1) We may construct an orthonormal basis $\{v_1, v_2, \ldots, v_n\}$ such that

$$A^*Av_i = \lambda_i v_i,$$

we may rearrange $\{v_1, v_2, \ldots, v_n\}$ such that $\lambda_j > 0$ for $j = 1, \ldots, r$ and $\lambda_j = 0$ for $j > r$. It is easy to see that

$$|Av_j|^2 = v_j^* A^* A v_j = \lambda_j v_j^* v_j = \lambda_j,$$

therefore let $\lambda_j = \sigma_j^2$, then $\frac{Av_j}{\sigma_j} = u_j$ is a unit vector for $j = 1, \ldots, r$. It is easy to verify

$$\langle u_j, u_k \rangle = \delta_{jk} \quad \text{for all } j, k \leq r.$$

Now we use the Gram–Schmidt theorem to extend $\{u_1, \ldots, u_n\}$ to an orthonormal basis $\{u_1, \ldots, u_m\}$, then we have

$$Av_i = \sigma_i u_i.$$

(2) We have,

$$V^*A^*AV = \begin{bmatrix} \Lambda & 0 \\ 0 & 0 \end{bmatrix},$$

where Λ is an $\ell \times \ell$ positive diagonal matrix. We take the first ℓ columns of V to form a matrix V_1, and the last $(n-\ell)$ columns to form a matrix V_2. Then we may rewrite the above system of equations as

$$V^*A^*AV = \begin{bmatrix} V_1^* \\ V_2^* \end{bmatrix} A^*A \begin{bmatrix} V_1 & V_2 \end{bmatrix} = \begin{bmatrix} V_1^*A^*AV_1 & V_1^*A^*AV_2 \\ V_2^*A^*AV_1 & V_2^*A^*AV_2 \end{bmatrix}$$

$$= \begin{bmatrix} \Lambda & 0 \\ 0 & 0 \end{bmatrix}.$$

Since V is unitary, we deduce that $V_1^*V_1 = I_\ell$, $V_2^*V_2 = I_{n-\ell}$. Define the $m \times \ell$ matrix $U_1 = AV_1\Lambda^{-1/2}$ where $\Lambda^{-1/2}$ is the diagonal matrix with the ith entry the $(-1/2)$th power of the corresponding term in Λ. Then we have

$$U_1 \Lambda^{1/2} V_1^* = AV_1 \Lambda^{-1/2} \Lambda^{1/2} V_1^* = A.$$

All are fine, except that U_1 is an $n \times \ell$ matrix, very far from a unitary matrix. However, we do have

$$U_1^* U_1 = \Lambda^{-1/2} V_1^* A^* A V_1 \Lambda^{-1/2} = \Lambda^{-1/2} \Lambda \Lambda^{-1/2} = I_\ell.$$

So that the columns of U_1 form an orthonormal set, therefore, we may complete it to a unitary matrix as

$$U = U_1 U_2.$$

Finally, it is easy to check that

$$A = U \Sigma V^*$$

where Σ is an $m \times n$ matrix with the upper left $n \times n$ submatrix Λ. ∎

Definition The diagonal entries σ_{ii} are called the *singular values* of A. The m columns of U are called left singular vectors of A, and the n column vectors of V are called the right singular vectors of A. ∎

Remark: The left singular vectors of A are the eigenvectors of AA^*. The right singular vectors of A are the eigenvectors of A^*A. The singular values of A are the square roots of the eigenvalues of both A^*A and AA^*. ∎

Lower Rank Approximation

Given a matrix A, we may find a *singular value decomposition* (SVD) of $A = U \Sigma V^*$. Let us only keep the largest r singular values σ_i. Thus we form a matrix $\tilde{\Sigma}$. We have a new matrix $\tilde{A} = U \tilde{\Sigma} V^*$ which is the rank r approximation of A (Eckart–Young theorem).

Image Processing

This approximation has a lot of applications in image processing, latent semantic analysis (information retrieval technique, etc.) and others. For instance, a satellite sending a picture to Earth. Say the picture has 1,000 "pixels" where each codes the color with 1,000 possible numbers. We may consider a $1,000 \times 1,000$ matrix A that consists of the data of a picture, and the satellite may send back the whole matrix. On the other hand we may consider the SVD of $A = U\Sigma V^*$ and keep only the 10% important singular values δ_i, i.e., the largest ones. And for each important δ_i, we keep the corresponding u_i and v_i such that

$$Av_i = \delta_i u_i.$$

Then the quality of the sent data which has only 200,000 numbers (as opposed to 1,000,000 numbers) produces a nice approximation of the original picture.

Latent Semantic Analysis (LSA)

We may analyze one article or several articles. In the first case we may separate the article into n paragraphs, in the second case we may treat each article as a unit. We list each paragraph in the first case or each article in the second case as columns of a matrix $A = (a_{ij})$. We list rows as the terms we deem as significant. Then we have a_{ij} as the measurement of the ith term in the jth column. We have a matrix A. Now we find its low rank approximation as the true meaning without the noise. This is the LSA. ∎

Definition (Polar Decomposition): Similar to writing a complex number c as $e^{i\theta}r$, we may write a square matrix A as $A = QS$ where Q is an orthogonal matrix, and S is symmetric positive semi-definite. In the complex case, any square matrix A can be written as $A = HU$ where H is hermitian and U is unitary.

It is easy to verify the above statements. Let us deal with the real case, where unitary matrices are orthogonal. Therefore $A = Q_1 \Sigma Q_2^T$, where Q_1, Q_2 are orthogonal, in the above equation of SVD, insert $I = Q_2^T Q_2$, and we have $A = (Q_1 Q_2^T)(Q_2 \Sigma Q_2^T) = QS$. The complex cases are similar.

One of the applications of the polar decomposition is in engineering. Let A denote a motion of material. Then we have the decomposition $A = QS$, where Q is rotation + reflection without any straining on the material, while S's eigenvalues cause the stretching or compression on the material. ∎

Exercises

(1) Find the SVD of the following matrix

$$\begin{bmatrix} 1 & 2 \\ 1 & 1 \\ 1 & 1 \end{bmatrix}.$$

(2) Find the SVD of the following matrix

$$\begin{bmatrix} 1 & 0 & 1 \\ 0 & 1 & 1 \\ 1 & 1 & 0 \end{bmatrix}.$$

(3) Let $A = [c_1, c_2, \ldots, c_n]$ be an $n \times n$ matrix with orthogonal column vectors. Find the SVD of it.

(4) Let a complex matrix A be written as HU where H is hermitian and U is unitary. Show that A is normal iff $H^2U = UH^2$.

(5) Let the SVD of A be $A = U\Sigma V^*$ where $\Sigma =$ a diagonal matrix. Let Σ^{-1} be the diagonal matrix which has σ^{-1} at the position of σ in the matrix Σ if and only if it is not zero. Define A^+ as $A^+ = V\Sigma^{-1}U^*$. Show that if the equation $Ax = b$ is solvable, then $x = A^+b$.

Chapter 6

Computational Linear Algebra

6.1 Numerical Model, Finite Difference Method and Finite Element Method

Originally, all mathematics started as mathematical models of the real world. In general, we have physical models, mathematical models, numerical models and computer models. We shall emphasize the values of linear algebra when it is applied as a mathematical model and a numerical model with the help of computers.

A physical model uses physical laws which is usually in the form of partial differential equations to describe a problem. A mathematical model uses the mathematical language. In a numerical model, we shall use a *discretization method*, such as the *theory of finite difference* or *finite element method*, to solve the mathematical model. In a computer model, we shall use computer programs to solve the numerical model. In this section, we will discuss approximation by: (1) a finite net of linear functions, (2) a finite set of known functions, (3) movable zones.

For instance, let us look at a mathematical model of an expanding economy.

von Neumann's Model of an Expanding Economy

The model of an expanding economy of von Neumann can be given as follows. Let us consider discrete time as monthly accounting or yearly accounting. We take $\mathbf{A}, \mathbf{B}$ to be $m \times n$ input and output real, non-negative matrices where the input matrix $\mathbf{A}$ includes the means of subsistence in the support of workers. The rows of matrices

247

$\mathbf{A}, \mathbf{B}$ are activity units (workshop, company, etc.), and columns are commodities. Then we have that on the ith row for the matrices $\mathbf{A} = (a_{ij}), \mathbf{B} = (b_{ij})$ with a_{ij} the amount of jth commodity required by ith unit and b_{ij} the amount of jth commodity produced by ith unit at time t. Let $\mathbf{q}^{(t)}$ be the m-dimensional vector of activity levels and at time t let $\mathbf{p}^{(t)}$ be the n-dimensional price vector at time t. Let us consider the following properties,

$$(\mathbf{q}^{(t)})^T \mathbf{B} \geq (\mathbf{q}^{(t+1)})^T \mathbf{A} \tag{1'}$$

$$\mathbf{B}\mathbf{p}^{(t)} \leq \mathbf{A}\mathbf{p}^{(t+1)} \tag{2'}$$

$$(\mathbf{q}^{(t)})^T \mathbf{B}\mathbf{p}^{(t+1)} = (\mathbf{q}^{(t+1)})^T \mathbf{A}\mathbf{p}^{(t+1)} \tag{3'}$$

$$(\mathbf{q}^{(t+1)})^T \mathbf{B}\mathbf{p}^{(t+1)} = (\mathbf{q}^{(t+1)})^T \mathbf{A}\mathbf{p}^{(t)} \tag{4'}$$

$$\mathbf{q}^{(t)} \geq \mathbf{0}, \text{ and } \mathbf{p}^{(t)} \geq \mathbf{0}. \tag{5'}$$

Condition (1') says that the input for a given period is not larger than the output of the previous period. (2') The total cost of each commodity produced from each activity unit does not exceed the total cost of commodity utilized for this unit of this activity at the next time, the profit is absent for all activity units. (3') The total cost of the industrial demand in the $(t+1)$th period equals to the total industrial income for commodities produced in the (t)th period. (4') The industrial expenses in the $(t+1)$th period (for (t)th period prices) equal to the total cost of outputs, i.e., the total amount of money is constant. (5') Natural requirements. Let us consider a uniform situation which has the following properties.

The number $\alpha = 1 + g$ is the expansion factor where g is the expansion or growth rate: we assume that all activity units grow with the same rate. We have $\beta = 1 + r$ that is the interest factor, where r is the rate of interest (or rate of profits): we assume that all prices are reduced with the same rate β^{-1},

$$\mathbf{q}^{(t+1)} = \alpha \mathbf{q}^{(t)} \quad \mathbf{p}^{(t+1)} = \beta^{-1}\mathbf{p}^{(t)}.$$

The axioms of the model are,

Definition (von Neumann's Model of Expanding Economy): Let us use the previously-defined notations. If the economy expanding model satisfies the following axioms,

$$\mathbf{q}^T\mathbf{B} \geq \alpha\mathbf{q}^T\mathbf{A} \tag{1}$$

$$\mathbf{Bp} \leq \beta\mathbf{Ap} \tag{2}$$

$$\mathbf{q}^T\mathbf{Bp} = \alpha\mathbf{q}^T\mathbf{Ap} \tag{3}$$

$$\mathbf{q}^T\mathbf{Bp} = \beta\mathbf{q}^T\mathbf{Ap} \tag{4}$$

$$\mathbf{q} \geq \mathbf{0}, \text{ and } \mathbf{p} \geq \mathbf{0}. \tag{5}$$

Then it is called the *von Neumann's model of expanding economy*. ∎

In order to demonstrate that for any pair of non-negative matrices $\mathbf{A}, \mathbf{B}$ there exist solutions for $\mathbf{q}, \mathbf{p}$, and for $\alpha \geq 0$ and $\beta \geq 0$, von Neumann assumes

$$\mathbf{A} + \mathbf{B} > 0$$

which implies that every process requires some positive amount of goods as an input or produces them in an output.

Let us consider an example of the von Neumann model of an expanding economy model.

Example 6.1: Let us consider a society with two industries: the food industry (F) (the first row) and the trade industry (T) (the second row). We will use the first column for food, and the second column for trade. We are given

$$A = \begin{bmatrix} 2 & 0 \\ 1 & 2 \end{bmatrix}, \quad B = \begin{bmatrix} 4 & 1 \\ 0 & 2 \end{bmatrix}.$$

It is easy to see that we may normalize the vectors $\mathbf{q}, \mathbf{p}$. Let us assume that $\alpha = \beta$. Then we have the inequalities,

$$\mathbf{q} \cdot (\mathbf{B} - \alpha\mathbf{A}) \geq \mathbf{0}, \quad (\mathbf{B} - \alpha\mathbf{A}) \cdot \mathbf{p} \leq \mathbf{0}.$$

If $\alpha > 2$, then clearly there is no solution. If $\alpha < 2$, then clearly there is no solution (see the Exercises). Therefore $\alpha = 2$ and $q = [1, 0]$, $p = [1, 0]^T$. So the whole society is concentrating in producing foods. ∎

Finite Net of Linear Functions

Note that in the expanding economy model it is already in the matrix form, hence a computer will help us with the solutions. In engineering, most problems are with partial differential equations. First, we have to reduce the problems to some algebraic equations. We need the theory of finite difference or *finite element method* for these reductions. Let us take simple examples to explain the *finite difference method* first,

Example 6.2: Let us consider the following problem. Let the solution of our problem satisfy the following differential equation,

$$\frac{d^2y}{(dx)^2} = -2.$$

Our boundary conditions are $y(0) = y(1) = 0$. Let us use the *finite difference method*. We further assume that $y(x) = 0$ for $x \leq 0$ or $x \geq 1$. Let us find the values of $y(x)$ at $x = 1/3$ and $2/3$. We use the convention $\Delta x = 1/3$, $\frac{dy}{dx} \approx \frac{\Delta y}{\Delta x} = \frac{y(x_0+\Delta x)-y(x_0)}{\Delta x}$, $\frac{d^2y}{dx^2} \approx \frac{\Delta^2 y}{\Delta x^2} = \frac{y(x_0-\Delta x)-2y(x_0)+y(x_0+\Delta x)}{\Delta x^2}$,

$$y'(0) = \frac{y(1/3) - y(-1/3)}{2\Delta x}, \quad y'(1/3) = \frac{y(2/3) - y(0)}{2\Delta x},$$

$$y'(2/3) = \frac{y(1) - y(1/3)}{2\Delta x}, \quad y'(1) = \frac{y(4/3) - y(2/3)}{2\Delta x},$$

$$y''(0) = \frac{y(1/3) + y(-1/3) - 2y(0)}{(\Delta x)^2},$$

$$y''(1/3) = \frac{y(2/3) + y(0) - 2y(1/3)}{(\Delta x)^2},$$

$$y''(2/3) = \frac{y(1/3) + y(1) - 2y(2/3)}{(\Delta x)^2},$$

$$y''(1) = \frac{y(2/3) + y(4/3) - 2y(1)}{(\Delta x)^2}.$$

Note that $y(0) = y(1) = 0$, $y''(1/3) = y''(2/3) = -2$. Let $a = y(1/3)$, $b = y(2/3)$. Then the differential equation $f'' = -2$ produces the following equations at $x = 1/3$ and $x = 2/3$,

$$b - 2a = -2(1/9)$$

$$a - 2b = -2(1/9).$$

We conclude $a = y(1/3) = 2/9$, $b = y(2/3) = 2/9$. While the exact solution of our differential equation with the values at the end points is $y = -x^2 + x$, which has values $2/9$ at the two points $x = 1/3, 2/3$.

The above can be put in matrix form $Ax = c$ as follows,

$$Ax = \begin{bmatrix} 2 & -1 & 0 & 0 \\ -1 & 2 & -1 & 0 \\ 0 & -1 & 2 & -1 \\ 0 & 0 & -1 & 2 \end{bmatrix} \begin{bmatrix} 0 \\ a \\ b \\ 0 \end{bmatrix} = \Delta x^2 \cdot \begin{bmatrix} -a/\Delta x^2 \\ 2 \\ 2 \\ -b/\Delta x^2 \end{bmatrix}.$$

In general, the matrix A is of the following form,

$$A = \begin{bmatrix} 2 & -1 & 0 & \cdots & \cdots & \cdots & 0 \\ -1 & 2 & -1 & 0 & \cdots & \cdots & 0 \\ 0 & -1 & 2 & -1 & 0 & \cdots & 0 \\ \cdots & \cdots & \cdots & \cdots & \cdots & \cdots & \cdots \\ \cdots & \cdots & \cdots & \cdots & \cdots & \cdots & \cdots \\ 0 & \cdots & \cdots & 0 & -1 & 2 & -1 \\ 0 & \cdots & \cdots & \cdots & 0 & -1 & 2 \end{bmatrix}.$$

It is an exercise to show that A is symmetric and positive-definite. ∎

In the following example, knowing the values of the function $y = f(x)$ at the nodes, we show how to approximate the function by *broken lines*,

Example 6.3: Suppose that values of the function $f(x)$ are given at $x_1 = 1/4$, $x_2 = 1/2$, $x_3 = 3/4$ as $f(x_i) = a_i$. We further define $x_0 = 0$, $f(0) = 0$, $x_4 = 1$, $f(1) = 0$. We shall use the *hat functions* g_i for $i = 1, 2, 3$ as follows,

$$g_i(x) = \begin{cases} 0 & \text{if } x \notin [x_{i-1}, x_{i+1}] \\ 4(x - x_{i-1}) & \text{if } x \in [x_{i-1}, x_i] \\ 4(x_{i+1} - x) & \text{if } x \in [x_i, x_{i+1}]. \end{cases}$$

Let $g(x) = \sum_{i=1}^{3} a_i g_i$. It is not hard to see that $f(x_i) = g(x_i)$ for $i = 1, 2, 3$ and $g(x)$ is a function of broken lines in the following graphs,

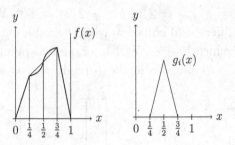

A Finite Set of Known Functions

We have the following example.

Example 6.4: Let us consider a ball thrown up at above the origin with unknown height along the x-axis, touched the ground at $x = 1$, and bounced up. We want to know the trajectory of the ball.

The trajectory of the ball obeys the following differential equation except at $x = 1$,

$$\frac{d^2y}{(dx)^2} = -g.$$

The behavior of the ball at $x = 1$ depends on: (1) what kind of ball is it? For instance, a basketball or a football, (2) the ground condition at $x = 1$. Those are difficult problems. An easy way is to separate the problem into two sections which are named *elements*: from $x = 0$ to $x = 1$, and from $x = 1$ to $x = 2$. Then we shall measure the values of y, at $x = 0, 2$, to solve two problems as two elements as follows. We use $1, x, x^2$ as the set of trial functions and $y = a_0 + a_1 x + a_2 x^2$, where a_i's are called *parameters* and are treated as variables. We have

$$2a_2 = -g \tag{1}$$

$$a_0 + a_1 + a_2 = 0 \tag{2}$$

$$a_0 = y_0 \quad (\text{or } a_0 + a_1(2) + a_2(2)^2 = y_2) \tag{3}$$

and then we assemble the two solutions to give a solution to the problem. ∎

In the preceding example there is a natural point $x = 1$ to separate the problem into two elements. In general we may just arbitrarily separate the whole domain into many small elements. Most engineering problems are presented in partial differential equations. Let us consider an approximation to a solution of a differential equation.

Example 6.5: Let us consider the following differential equation,

$$\frac{d^2y}{dx^2} + x\frac{dy}{dx} = x + \frac{1}{1+x} \quad \text{where } 0 \le x \le 1.$$

Let us use the trial functions $\{\sin(\pi x), \cos(\pi x), \sin(2\pi x), \cos(2\pi x)\}$. Let an approximate function f be written as a linear combination of these four functions as

$$f(x) = a_1\sin(\pi x) + a_2\cos(\pi x) + a_3\sin(2\pi x) + a_4\cos(2\pi x).$$

We are looking for function $f(x)$ which satisfies the differential equation at $x = 1/6, 1/4, 3/4, 5/6$. We have

$$-a_1(\pi)^2(1/2) - a_2(\pi)^2(\sqrt{3})/2 - 4a_3(\pi)^2(\sqrt{3}/2) - 4a_4(\pi)^2(1/2)$$
$$+ (1/6)(a_1\pi(1/2) - a_2\pi(\sqrt{3}/2) + a_32\pi(\sqrt{3}/2) - a_42\pi(1/2))$$
$$= 43/42,$$
$$-a_1(\pi)^2(\sqrt{2}/2) - a_2(\pi)^2(\sqrt{2})/2 - 4a_3(\pi)^2$$
$$+ (1/4)(a_1\pi(\sqrt{2}/2) - a_2\pi(\sqrt{2}/2) + a_42\pi) = 21/20,$$
$$-a_1(\pi)^2(\sqrt{2}/2)r + a_2(\pi)^2(\sqrt{2})/2 + 4a_3(\pi)^2$$
$$+ (3/4)(-a_1\pi(\sqrt{2}/2) + a_2\pi(\sqrt{2}/2) + a_42\pi) = 37/28,$$
$$-a_1(\pi)^2(\sqrt{3}/2) + a_2(\pi)^21/2 - 4a_3(\pi)^2(1/2) - 4a_4(\pi)^2(\sqrt{3}/2)$$
$$+ (5/6)(-a_1\pi(1/2) + a_2\pi(\sqrt{3}/2) + a_42\pi(1/2)) = 91/66.$$

We solve the system of linear equations by using a computer program (see the Exercises), and we have an approximate solution of our differential problem this way. ∎

Movable Zones

Example 6.6: Let us have a system of springs arranged as follows, where u_i is the displacement of the ith node for $i = 1, 2, 3$. The spring constants are k_i for $i = 1, 2$. The complicated spring system can be separated into two elements, with each one shown as the following diagrams,

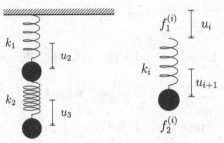

We have the following relations for $i = 1, 2$,

$$k_i(u_i - u_{i+1}) = f_1^{(i)}$$
$$k_i(u_{i+1} - u_i) = f_2^{(i)}.$$

Let us assume that $u_1 = 0$, i.e., the whole system is hanging on a fixed place. We shall assemble the system in the following system of equations,

$$k_1(u_1 - u_2) = f_1^{(1)}$$
$$k_1(u_2 - u_1) = f_2^{(1)}$$
$$k_2(u_2 - u_3) = f_1^{(2)}$$
$$k_2(u_3 - u_2) = f_2^{(2)}.$$

We consider the total forces at the node points as $F_1 = f_1^{(1)}$, $F_2 = f_1^{(2)} + f_2^{(1)}$, $F_3 = f_2^{(2)}$. Then the system can be written as

$$\begin{bmatrix} k_1 & -k_1 & 0 \\ -k_1 & k_1 + k_2 & -k_2 \\ 0 & -k_2 & k_2 \end{bmatrix} \begin{bmatrix} 0 \\ u_2 \\ u_3 \end{bmatrix} = \begin{bmatrix} F_1 \\ F_2 \\ F_3 \end{bmatrix}.$$

Let us give the problem some numerical values. Let $k_1 = 50 \, \text{kg/cm}$ and $k_2 = 75 \, \text{kg/cm}$. Let $F_2 = F_3 = 75 \, \text{kg} \cdot \text{m/sec}^2$. Then we have the following system of equations,

$$-50u_2 = F_1$$
$$125u_2 - 75u_3 = 75$$
$$-75u_2 + 75u_3 = 75.$$

We may solve the above system and get $u_2 = 3$, $u_3 = 4$, $F_1 = -150$.

■

Depending on the problem, we have a set of trial functions $\{u_i\}$, and we have a solution

$$f = \sum_i^n a_i u_i$$

where the a_i's are called the *parameters*. We will substitute the function f to the differential equations, to produce a set of algebraic equations of the parameters a_i. Then we use a computer to help us to solve the set of algebraic equations approximately. If our original differential equations are linear, then we have a system of linear equations. We may use the theory of matrices to handle them or we have movable zones.

In modern times, there are problems of weather forecasting, pollution, water course management, space travel and nuclear applications. Usually, we set up stations — weather stations and gauging stations and measure stations — to collect data for smaller zones and solve the simple problems for smaller zones, and then assemble all solutions of the smaller zones to a global solution.

Exercises

(1) Finish the discussion of **Example 6.1**.
(2) Show the $n \times n$ matrix A in Example 6.2 is symmetric and positive-definite.
(3) Let us consider the following problem. Let the solution of our problem satisfy the following differential equation,

$$\frac{d^2y}{(dx)^2} = -2.$$

Our boundary conditions are $y(0) = y(1) = 0$. Let us use the *theory of finite difference method*. Let us find the value of $y(x)$ at $x = 1/2$. We use the convention $y''(1/2) = \frac{y(1)+y(0)-2y(1/2)}{(\Delta x)^2}$ (cf. **Example 6.2**).
(4) Finish the discussion of **Example 6.5**.

(5) Compute Example 6.6 with a different data, $k_1 = 10\,\text{kg/cm}$, $k_2 = 20\,\text{kg/cm}$ and $F_1 = F_2 = 30\,\text{kg}\cdot\text{m/sec}^2$. Find u_2, u_4 and F_1.

(6) Find g in Example 6.3 with $x_1 = 1$, $x_2 = 2$, $x_3 = 1.5$.

6.2   Power Method. Perron–Frobenius Theorem. Google Search

Power Method

Given a square complex matrix A. Let us assume that its eigenvalues $|\lambda_1| > |\lambda_2| \geq |\lambda_3| \geq \cdots \geq |\lambda_s|$. We further assume that the multiplicity of λ_1 is 1. We do not know the precise values of λ_j. To find them, we have to solve the characteristic polynomial $\chi_A(x)$ which is difficult to find and solve (we will deal with the problem in the next section). Sometimes from theoretic arguments, we know that the eigenvalues satisfy the above inequality, we wish to use a fast power method to find λ_1 and its associated eigenvector w_1, i.e., $Aw_1 = \lambda_1 w_1$.

Let us assume that we find the Jördan[1] canonical form of A and the basis $\{w_1, \ldots, w_n\}$, with Jördan blocks $\{w_i, \ldots, w_j\}$ such that

$$Aw_i = \lambda_k w_i$$
$$Aw_{i+1} = \lambda_k w_{i+1} + w_i$$
$$\cdots\cdots$$
$$Aw_j = \lambda_k w_j + w_{j-1}.$$

Here we assume that the Jördan block of λ_1 is unique and of length 1. Then it is easy to see that

$$A^m w_i = (\lambda_k)^m w_i$$
$$A^m w_{i+1} = (\lambda_k)^m w_{i+1} + C_1^m (\lambda_k)^{m-1} w_i + \cdots$$
$$\cdots\cdots$$
$$A^m w_j = (\lambda_k)^m w_j + C_1^m (\lambda_k)^{m-1} w_{j-1} + \cdots.$$

Let $v = b_1 w_1 + b_2 w_2 + \cdots + b_n w_n$. For most vectors v, it is likely that all b_i are non-zeroes. There is a danger for $A^m v = b_1 \lambda_1^m v_1 + \cdots$

---

[1] Jördan, C. French mathematician. 1838–1922.

to blow-up, if $|\lambda_1| > 1$. On the other hand, $A^m v = b_1 \lambda_1^m v_1 + \cdots$ may blow-down to zero, if $|\lambda_1| < 1$. Those two situations are bad, and we avoid them by restricting the sizes of $\{v_i\}$ where $v_1 = v$, $v_2 = a_2 A v_1, \ldots, v_m = a_m A v_{m-1} = a_m A^{m-1} v_1, \ldots$ by a suitably selected sequence $a_2, \ldots$. For instance, we require that $1/2 < |v_m| < 1$. Or the maximal coordinates v_{mi} of v_m satisfies $1/2 < v_{mi} < 1$ etc. Then it is not hard to see

$$\lim_{m \mapsto \infty} \left(\prod_{i=2}^{i=m} a_i \right)^{1/(m-1)} = \lambda_1.$$

We may replace A by A/λ_1, then the new largest eigenvalue will be 1, and $1 > |\lambda_2| \geq |\lambda_3| \geq \cdots$. We have the following proposition which describes the first eigenvector w_1.

Proposition 6.2.1: *Let us use the preceding notations and assumptions. Then we have*

$$\lim_{m \mapsto \infty} A^m v = w_1.$$

Proof. Note that under our assumption of $\lambda_1 = 1$ and with multiplicity 1, all $\lambda_r < \epsilon$ for $r > 1$, the coefficients of the expressions of $A^m w_{i+s}$ for $1 < i \leq i + s \leq j$ are either $< \epsilon^m$ or $< \epsilon^{m-s} C_r^m$ where $s, r < n$. It is easy to see that all coefficients go to zero except the first one, and $A^m v$ goes to w_1. ∎

Perron–Frobenius Theorem

Can all the above conditions on the matrix A be satisfied in some interesting cases? We have an interesting case covered by the following *Perron*[2]–*Frobenius*[3] theorem. To begin with, we will define,

Definition A real matrix $A = (a_{ij})$ is said to be positive, if $a_{ij} > 0$ for all i, j. The matrix A is said to be non-negative, if $a_{ij} \geq 0$ for all i, j. ∎

---

[2]Perron, O. German mathematician. 1880–1975.
[3]Frobenius, F. German mathematician. 1849–1917.

We shall have some notations,

Notations: Let $M = (m_{tj})$ be a square matrix. Let $\sigma(M)$ be the set of eigenvalues of M. For any complex or real matrix M, let $\rho(M)$ be the largest absolute value of all eigenvalues. Let $|M| = (|m_{tj}|)$. Especially, a complex vector v can be considered as a matrix, then $|v|$ is defined for any complex vector v. ∎

We have the following important **Perron Theorem** (1907) for positive square matrices, which is later generalized to the **Perron–Frobenius Theorem** for some non-negative matrices by Frobenius (1912). In the following statements, we only use the matrix A to act from left as Av. Similar statements also work for the matrix A to act from right as $v^T A$. Since both ways share the same characteristic polynomial, the eigenvalues from both ways must be the same. The only thing we have to pay attention to is that for a fixed eigenvalue λ_r, the left eigenspace associated with λ_r may be different from the conjugate of the right eigenspace associated with the same eigenvalue λ_r.

Theorem 6.2.2 (Perron Theorem): *Let $M = (m_{tj})$ be a real positive square matrix. Then we have,*

(1) *Let $r = \rho(M)$, then $r > 0$ and $r \in \sigma(M)$. Further, $r >$ the absolute value of any other eigenvalue.*

(2) *The eigenvalue r has a* **positive** *vector v as its associated eigenvector. Any other eigenvalue has* **no** *non-negative associated eigenvector.*

(3) *The eigenvalue r is a* **simple root** *of the characteristic polynomial of M.*

Proof. (1) If $r = 0$, it follows from the Jördan canonical form of M that M is nilpotent, which is not true under our assumption. Therefore $r > 0$. Let us define a set $S = \{s : Mv \geq sv \text{ for some } v \text{ non-negative and} \neq 0\}$. Clearly the set S is non-empty and has some positive elements. For instance, let $v = [1, 1, \ldots, 1]^T$ and $m = \min(\{\sum_j m_{tj}\})$. Then $m > 0$ and $Mv \geq mv$. Let $s = \sup \overline{S}$. We claim $s \in \sigma(M)$, i.e., s is an eigenvalue.

By considering only the vector v such that $||v|| = 1$ for any norm $|| \; ||$, we have a compact set. Then we pick a subsequence v_i such that $Mv_i \geq \lambda_i v_i$ where $\lambda_i :\mapsto s$ with the further property that $v_i :\mapsto v$, and we have $Mv \geq sv$. If it is an equality, then s is an eigenvalue. Let us assume that it is not an equality (which does not mean that it is a straight inequality). Let $u = Mv - sv$. Then u is non-negative and non-zero. Therefore $0 < Mu = M(Mv) - s(Mv)$. Let $w = Mv$, we have $Mw > sw$. We may increase the value of s by ϵ, and the inequality would still work. It contradicts the assumption that s is the maximal possible one. Hence we show that s is an eigenvalue.

Let λ be any other eigenvalue (i.e., $\lambda \neq s$) with associated eigenvector z. Then we have $Mz = \lambda z$. Take the absolute values on both sides, we have $M|z| \geq |Mz| = |\lambda||z|$. Therefore $|\lambda| \in S$ and $|\lambda| \leq s$. Suppose $|\lambda| = s$ and we must have $M|z| = |\lambda||z|$ (otherwise, by a previous argument, s can be increased, a contradiction). Hence $M|z| = |Mz|$, $\sum_j m_{tj}|z_j| = |\sum_j m_{tj}z_j|$. The preceding equation happens only if $z_j = e^{\theta i}|z_j|$ for all j. Furthermore, we have $Me^{\theta i}|z| = \lambda e^{\theta i}|z|$, thus $M|z| = \lambda|z|$. Therefore, $\lambda = |\lambda|$. We conclude $\lambda = s$ which contradicts with the assumption that λ is distinct from s.

(2) We have $Mv = rv$ for some non-negative v. The left side Mv is a positive vector, hence the right side rv is a positive vector, thus v is positive. Similarly M has a positive left eigenvector w^T associated with r. Let u be an (right) non-negative eigenvector associated with another eigenvalue λ. Then we have

$$rw^T u = w^T M u = \lambda w^T u \quad \text{and} \quad r \neq \lambda$$

hence

$$w^T u = 0.$$

Since w^T is positive, then $w^T u$ is positive. A contradiction. Therefore (2) is verified.

(3) Let us make some simplifications. We replace M by M/r, and assume $\rho(M) = 1$. Furthermore, we may select the basis so that M is in the Jördan canonical form J. In other words, we may assume that

$$J = P^{-1}MP = \begin{bmatrix} J_1 & 0 & 0 & 0 & 0 \\ 0 & J_2 & 0 & 0 & 0 \\ 0 & 0 & \cdot & 0 & 0 \\ \cdot & \cdot & \cdot & \cdot & \cdot \\ 0 & 0 & 0 & 0 & J_h \end{bmatrix},$$

where J_h's are Jördan blocks. It is easy to see that

$$J^k = P^{-1}M^kP = \begin{bmatrix} J_1^k & 0 & 0 & 0 & 0 \\ 0 & J_2^k & 0 & 0 & 0 \\ 0 & 0 & \cdot & 0 & 0 \\ \cdot & \cdot & \cdot & \cdot & \cdot \\ 0 & 0 & 0 & 0 & J_h^k \end{bmatrix}.$$

We assume that J_1 is associated with eigenvalue 1 with the largest size m. We claim that $m = 1$. Let us assume that $m > 1$. Note that we have $M = PJP^{-1}$. Let $v = [v_1, v_2, \ldots, v_n]^T$ be a positive eigenvector associated with 1. Clearly we have $M^k v = v$ for all integer $k \geq 1$. Note that $\min_t\{v_t\} = c > 0$. Let $|| \ ||_\infty$ be the norm defined as

$$||B||_\infty = \max\left\{\sum_j |b_{tj}|\right\}.$$

Then $||v||_\infty = \max\{|v_t|\}$. Let us compute $||J_1^k||_\infty$ for J_1 associated with eigenvalue 1 and of size $m > 1$. We have

$$J_1^k = \begin{bmatrix} 1 & 0 & 0 & 0 & 0 \\ k & 1 & 0 & 0 & 0 \\ \cdot & k & 1 & 0 & 0 \\ \cdot & \cdot & \cdot & \cdot & \cdot \\ \cdot & \cdot & \cdot & k & 1 \end{bmatrix}.$$

It follows easily that $||J^k||_\infty \mapsto \infty$ as $k \mapsto \infty$. We have $||J^k||_\infty = ||P^{-1}M^kP||_\infty \leq ||P^{-1}||_\infty ||M^k||_\infty ||P||_\infty$, which means

$$||M^k||_\infty \geq \frac{||J^k||_\infty}{||P^{-1}||_\infty ||P||_\infty} \mapsto \infty.$$

Let $M^k = (m_{tj}^{(k)})$. Recall that $M^k v = v$. Then we have

$$||v||_\infty = \max\{|v_t|\} = ||M^k v||_\infty = \max\left\{\sum_j m_{tj}^{(k)} |v_j|\right\}$$

$$\geq \max\left\{\sum_j m_{tj}^{(k)}\right\} \min\{|v_t|\} = ||M^k||_\infty c \mapsto \infty$$

which is impossible.

We conclude that all Jördan blocks associated with 1 are 1×1. We want to show that there is only one Jördan block associated with 1. Suppose that there are two. Let u, v be linearly independent eigenvectors. We may assume that tth component v_t of v is not zero. Let $w = u - (u_t/v_t)v$. Then w is also an eigenvector associated with 1. We have $Mw = w$. Taking absolute values on both sides, we have $|M||w| \geq |Mw| = |w|$. If the inequality is not equal, we may find a larger eigenvalue, which is impossible. Therefore, we must have $M|w| = |w|$. Since the left-hand side is positive, so must the right-hand side be, which is impossible as we know that the tth component is zero. ∎

Definition The number r in the preceding theorem is called the *Perron root*. The unique eigenvector $v = [v_1, v_2, \ldots, v_n]^T$ associated with r such that $\sum_t v_t = 1$ is called the *Perron vector*. ∎

We will mention the generalization made by Frobenius for the completeness. Frobenius found a class of non-negative matrices such that the statements of the Perron Theorem are valid. A square matrix is said to be *irreducible* if the following definition is satisfied.

Definition A matrix $A_{n \times n}$ is said to be *irreducible* iff there is no permutation matrix P such that

$$P^{-1}AP = \begin{bmatrix} X & Y \\ 0 & Z \end{bmatrix}$$

where X, Z are smaller square matrices. ∎

Definition A matrix $A_{n \times n}$ is said to be *primitive* iff A^k is positive for some positive integer k. ∎

Proposition 6.2.3: *Every primitive matrix is irreducible.*

Proof. It is sufficient to prove that the kth power of a reducible (i.e., not irreducible) matrix A is reducible, which is trivial. ∎

We will state the following proposition without giving a proof. Interested readers may consult Chapter 8 of C. Meyer's book

Matrix Analysis and Applied Linear Algebra, SIAM, http://www. matrixanalysis.com/Chapter8.pdf.

Proposition 6.2.4: *An irreducible matrix $A = (a_{tj})$ is a primitive matrix iff a_{tt} is not 0 for some t.*　　　　　　　■

We have the following theorem,

Theorem 6.2.5 (Perron–Frobenius Theorem): *Let $M = (m_{tj})$ be a real irreducible non-negative square matrix. Then we have,*

(1) *Let $r = \rho(M)$, then $r > 0$ and $r \in \sigma(M)$. We have r as a simple root of the characteristic polynomial.*

(2) *The eigenvalue r has a positive vector v as its associated eigenvector. If an eigenvalue λ is with $\lambda \neq r$, then it has no non-negative associated eigenvector.*

(3) *Furthermore if M is primitive, then $r = \rho(M)$ is the only eigenvalue which has the largest absolute value.*

Proof. We shall only prove the case that A is primitive in this book. For the general case, see Chapter 8 of C. Meyer's book *Matrix Analysis and Applied Linear Algebra*, SIAM, http://www. matrixanalysis.com/Chapter8.pdf. Applying the Perron Theorem to M^k, we have the above statement for M^k. We have to establish a relation between the eigenvalues of M and M^k. Let us select a suitable basis for the vector space so that M is expressed in its Jördan canonical form. Then it is easy to see that the diagonal items are the eigenvalues $\{\lambda_t\}$ of M. Furthermore, the diagonals of M^k will be $\{\lambda_t^k\}$. It follows that the characteristic polynomial of M^k is $\prod(\lambda - \lambda_t^k)^{n_t}$, which establishes a bijective correspondent between the set of eigenvalues of $\{\lambda_t\}$ of M and the eigenvalues $\{\lambda_t^k\}$ of M^k.

The only thing we have to prove is that there is a positive eigenvector v associated with the largest eigenvalue r. Pass to M^k, we know that v is an eigenvector associated with the largest eigenvalue r^k of M^k. Hence it must be either positive or $e^{i\theta}v$ is positive, for some θ. So r has a positive associated eigenvector.　　　■

We have the following definition,

Definition A non-negative square matrix $A = (a_{tj})$ is called a *stochastic matrix* iff every column adds to 1, i.e.,

$$\sum_t a_{tj} = 1 \quad \text{for all } j.$$

A vector u is called a *stochastic vector* iff it is a stochastic matrix. ∎

We have the following proposition for a stochastic matrix.

Proposition 6.2.6: *Let $A = (a_{tj})$ be an irreducible stochastic matrix. Then 1 is its largest eigenvalue.*

Proof. Let $v = [1, 1, \ldots, 1]$. Then clearly $vA = v$. Therefore 1 is an eigenvalue of A and with a (left) positive eigenvector v. It follows from the Perron–Frobenius Theorem that 1 is the largest eigenvalue of A for all vectors from the right. ∎

Input-output Model

The model was invented by W. Leontief[4] in the 1930's. It is assumed that in an economy, there are n sectors. Let x_i = total output of sector i for $i = 1, 2, \ldots, n$, b_i = total final demand (external demand) for the products of sector i for $i = 1, 2, \ldots, n$, and a_{ij} = input of product of sector i required by sector j to produce one unit of its product for $i, j = 1, 2, \ldots, n$. Note that if we assume that the relation between input and output is linear, (while there are many laws in economy, like the *diminished return*, that tell us it is not linear, in any case even if linearity is not true, linearity is a good approximation),

---

[4]Leontief, W. Russian-born American economist. 1973 Nobel Prize winner. 1906–1999.

then we have

$$x_i = \sum a_{ij}x_j + b_i.$$

Or we may write in the matrix form as follows,

$$x = Ax + b.$$

Or

$$(I - A)x = b, \quad x = (I - A)^{-1}b.$$

The matrix $(I - A)^{-1}$ is all important in Leontief's model. We further require that it is non-negative. Then we have that

$$(I - A)^{-1} = I + A + A^2 + \cdots + A^n + \cdots .$$

Let $\lambda =$ the Perron root of A. Then one can show that the above power series converges iff $\lambda < 1$. Hence it will be thus assumed. There is a proposition (Hawkins–Simon condition) that if the principal minors of $I - A$ are all positive, the required output vector x is non-negative.

Example 6.7: Let A be the following matrix,

$$A = \begin{bmatrix} 0 & 0.5 & 0.55 \\ 0.25 & 0.05 & 0.1 \\ 0.25 & 0.05 & 0 \end{bmatrix}.$$

Then it is easy to see that

$$I - A = \begin{bmatrix} 1 & -0.5 & -0.55 \\ -0.25 & 0.95 & -0.1 \\ -0.25 & -0.05 & 1 \end{bmatrix}.$$

The principal minors are $1, 0.825, 0.70375$. Therefore the Hawkins–Simon condition is satisfied. Direct computation shows that

$$(I - A)^{-1} = \begin{bmatrix} 1.4104 & 0.7873 & 0.8545 \\ 0.4104 & 1.2873 & 0.3545 \\ 0.3731 & 0.2612 & 1.2323 \end{bmatrix}.$$

and the Perron root of $I - A$ is 0.7773.

Remark: Leontief's work is significant. Leontief separated the US economy into 500 sectors and tried to apply his theory. The results were unclear. ∎

Leontief's theory can be modified into a linear programming problem (see Section 6.5) as finding x such that

$$\min \sum c_i x_i$$

subject to

$$(I - A)x \geq b, \quad x \geq 0.$$

The Ranking Problem of a Chess Tournament

Ranking of Chess Players

Let us consider a round robin chess tournament for six players $P_1, P_2, P_3, P_4, P_5, P_6$. We use their records to form the following matrix, where $a_{ij} = 1$ if P_i defeats P_j, and in this situation $a_{ji} = 0$. If P_i ties with P_j, then $a_{ij} = a_{ji} = 0.5$. We always take $a_{ii} = 0.5$. Let the results of this tournament be represented by the following matrix of the relative strengths,

$$A = \begin{bmatrix} 0.5 & 1 & 1 & 0 & 1 & 1 \\ 0 & 0.5 & 0 & 1 & 1 & 0 \\ 0 & 1 & 0.5 & 1 & 1 & 1 \\ 1 & 0 & 0 & 0.5 & 0 & 0 \\ 0 & 0 & 0 & 1 & 0.5 & 1 \\ 0 & 1 & 0 & 1 & 0 & 0.5 \end{bmatrix}.$$

At the beginning we are ignorant of the relative strength of each player and let v_1 be the vector $[1, 1, 1, 1, 1, 1]^T$ to assume their equal strengths. Then we may take v_2 as follows to indicate their temporary relative strengths after the tournament,

$$v_2 = Av_1 = \begin{bmatrix} 0.5 & 1 & 1 & 0 & 1 & 1 \\ 0 & 0.5 & 0 & 1 & 1 & 0 \\ 0 & 1 & 0.5 & 1 & 1 & 1 \\ 1 & 0 & 0 & 0.5 & 0 & 0 \\ 0 & 0 & 0 & 1 & 0.5 & 1 \\ 0 & 1 & 0 & 1 & 0 & 0.5 \end{bmatrix} \begin{bmatrix} 1 \\ 1 \\ 1 \\ 1 \\ 1 \\ 1 \end{bmatrix} = \begin{bmatrix} 4.5 \\ 2.5 \\ 4.5 \\ 1.5 \\ 2.5 \\ 2.5 \end{bmatrix}.$$

We may take v_2 as the ranking vector, and conclude the order of players as $P_1 = P_3 > P_2 = P_5 = P_6 > P_4$. This kind of ranking is used in newspapers for college football teams and Olympic Games. However, there are two problems in the present tournament: (1) The first player and the third player have the same score 4.5. Does the fact that the first player defeated the third player at their head-on contest mean something? (2) The fourth player defeated a strong player (the first player), does it mean something?

To amend the situation, we shall consider the temporary relative strengths of all players as indicated by temporary vector $v_2 = Av_1$, and adopt the rule that: (1) if player P_i defeats player P_j, then P_i adds P_j's strength to his/her own strength, and (2) if player P_i ties with player P_j, then P_i adds half of P_j's strength to his/her own strength. In the comparison of all college football teams in the US, this step is called *consider the schedules* or *consider the strength of the opponents*. We then compute $v_3 = Av_2 = A^2v_2$. We have

$$v_3 = Av_2 = A^2v_1 = \begin{bmatrix} 0.25 & 3 & 1 & 4 & 3 & 3 \\ 1 & 0.25 & 0 & 2 & 1 & 1 \\ 1 & 2 & 0.25 & 4 & 2 & 2 \\ 1 & 1 & 1 & 0.25 & 1 & 1 \\ 1 & 1 & 0 & 2 & 0.25 & 1 \\ 1 & 1 & 0 & 2 & 1 & 0.25 \end{bmatrix} \begin{bmatrix} 1 \\ 1 \\ 1 \\ 1 \\ 1 \\ 1 \end{bmatrix}$$

$$= \begin{bmatrix} 14.25 \\ 5.25 \\ 11.25 \\ 5.25 \\ 5.25 \\ 5.25 \end{bmatrix}.$$

We may now use v_3 as the new temporary relative strength vector and use it as the ranking vector. At this iteration, the first player is pulled apart from the third player, and the fourth player gets an equal rank with the remaining players. Note that this is the way that the "computer ranking" works for the college football teams in the US. If we carry the argument one step further, we find $v_4 = Av_3 = A^3v_1$

as follows,

$$v_4 = Av_3 = A^2 v_2 = A^3 v_1$$

$$= \begin{bmatrix} 4.125 & 5.75 & 0.75 & 12 & 5.75 & 5.75 \\ 2.5 & 2.125 & 1 & 3.25 & 1.75 & 2.5 \\ 4.5 & 4.25 & 1.125 & 8.25 & 4.25 & 4.25 \\ 0.75 & 3.5 & 1.5 & 4.125 & 3.5 & 3.5 \\ 2.5 & 2.5 & 1 & 3.25 & 2.125 & 1.75 \\ 2.5 & 1.75 & 1 & 3.25 & 2.5 & 2.125 \end{bmatrix} \begin{bmatrix} 1 \\ 1 \\ 1 \\ 1 \\ 1 \\ 1 \end{bmatrix}$$

$$\approx \begin{bmatrix} 34 \\ 13 \\ 27 \\ 17 \\ 13 \\ 13 \end{bmatrix}.$$

We can show that if we carry on the argument further, the order $P_1 > P_3 > P_4 > P_2 = P_5 = P_6$ will remain the same. We may conclude that it is the final order.

Note that A is a *non-negative matrix* and A^k is a *positive matrix* for $k \geq 3$.

Google Rank and PageRank

Note that any non-negative matrix A without a zero column can be *normalized* to a stochastic matrix, i.e., replace every column by the new column by dividing the old column by its sum. Now we have enough mathematical background to discuss the Google search. (1) Once the user inputs some keywords, the Google Engine collects all websites $W_1, W_2, \ldots, W_n$ (in general, it assumes that $n \leq 10,000$) with those keywords (this is common to all search engines, and we will not discuss it). (2) The Google Engine ranks all collected websites. For this purpose, let us proceed as follows.

Let us form a matrix $A = (a_{tj})$, where a_{tj} = the number of references from website W_j to website W_t (it is our way to count *citations*). Then we normalize every non-zero column by replacing a_{tj} by $a_{tj}/\sum_t a_{tj}$ and still call the matrix $A = [a_1, a_2, \ldots, a_n]$, where a_j are the column vectors of A. We now define a new matrix $B = [b_1, b_2, \ldots, b_n]$ where b_j are the column vectors of B as follows: if $a_j = 0$, then $b_j = [1/n, 1/n, \ldots, 1/n]^T$; if $a_j \neq 0$, then $b_j = 0$. Then $A + B$ will be a stochastic matrix. The trouble is that $A + B$ is only non-negative, while it may not be primitive. Anyway, it is troublesome to check if it is primitive. We may speculate the way Google uses to proceed. (1) Google creates another $n \times n$ stochastic matrix Q named *taste*, where it is a random matrix and takes random stochastic column vectors, and formulates $C = x(A + B) + (1 - x)Q$ where $0 < x < 1$. Then C is primitive for x nearing 1 (in fact, Sergey Brin and Larry Page state that it is good to take $x = 0.85$) and we know that the largest eigenvalue is 1. Now the work is to find the all-important eigenvector (Perron vector) v associated with 1. The values of components of v will give us the ranking of the websites. We shall find the Perron vector v using the following power method.

Let us use the notations of a basis $\{v = w_1, w_2, \ldots, w_n\}$ such that

$$Cw_1 = w_1 \quad \text{or} \quad Cw_t = \lambda_j w_t \quad \text{or} \quad Cw_t = \lambda_j w_t + w_{t+1}$$

$$\text{where } |\lambda_j| < 1.$$

Then we have

$$C^2 w_1 = w_1 \quad \text{or} \quad C^2 w_t = \lambda_j^2 w_t \quad \text{or}$$

$$C^2 w_t = \lambda_j^2 w_t + c_1^2 \lambda_j w_{t+1} + \lambda_j w_{t+2} \quad \text{where } |\lambda_j| < 1.$$

It is easy to see that

$$C^k w_1 = w_1 \quad \text{or} \quad C^k w_t = \lambda_j^k w_t \quad \text{or}$$

$$C^k w_t = \lambda_j^k w_t + c_1^k \lambda_j^{k-1} w_{t+1} + \cdots \quad \text{where } |\lambda_j| < 1.$$

We can easily conclude that

$$\lim_{k \mapsto \infty} C^k w_1 = w_1 \quad \lim_{k \mapsto \infty} C^k w_t = 0 \quad \text{for all } t > 1.$$

Let us randomly pick up a vector $u = a_1 w_1 + w_2 v_2 + \cdots + a_n w_n$ with $a_1 \neq 0$. Then by the above argument, we have $\lim_{k \mapsto \infty} C^k u = a_1 w_1$.

After normalizing $a_1 w_1$, we find the Perron vector $w_1 = v$. Hence we have the Google ranking.

According to Larry Page, starting with a random vector u, it suffices to take $k = 50$ to get a good approximation of $a_1 w_1$ for the ranking purpose. There are many fast ways of computing C^k in computational mathematics.

Or (2) we take a matrix O such that $o_{tj} = 1/n$ for all t, j, a vector $u = (1/n, 1/n, \ldots, 1/n)^T$, and $D = x(A+B) + (1-x)O$. Note that O is stochastic. Similarly, we take $x = 0.85$. Note that u is a stochastic vector, and Dv is stochastic if D, v are stochastic. Let $u_1 = u$ and inductively

$$u_{m+1} = Du_m = x(A + B)u_m + (1 - x)Ou_m$$

$$= xAu_m + xBu_m + (1 - x)Ou_m = xAu_m + xBu_m + (1 - x)u$$

since it is easy to see that inductively all u_m's are stochastic. In the preceding expression, the first term xAu_m is with a *sparse* matrix A where most entries are zero. It is likely that there are only n non-zero entries (we assume that among the 10,000 websites, there are only about 10,000 cross references). So to compute it out, first we pre-compute xA and xB (it will be used throughout the computing process), and likely we need only n multiplications for the computation of xA, since xB is a matrix with 0 or x/n (for those columns corresponding to 0 columns of A only), we need only one multiplication. We compute $(xA)u_m$ next. With xA known, then we only need n multiplications. The second term xBu_m is with the columns of B either 0 or $u = [1/n, \ldots, 1/n]^T$. It is easy to see that $xBu_m = (a, a, \ldots, a)^T$, where $a = x/n(\sum_{t_j} u_{t_j})$ where t_j runs through only those 0 columns of A and u_{t_j} are the entries of u_m which are corresponding to t_j and there are at most one multiplication needed. To find u_{m+1} from u_m, we need $n + 1$ multiplications plus $n + 1$ multiplications of pre-computations (which can be used again and again). Note that the number of multiplications is linear in n. For instance let $n = 10{,}000$ and $m = 50$. Then we need 510,051 multiplications to finish the Google ranking. It is a fast job.

Note that the value of x does not affect the speed of computation, we may take x to be as close to 1 as possible, say $x = 0.99$. ∎

Multi-criteria Ranking Problem

Let us assume that we have m decision alternatives $k = 1, \ldots, m$ and t decision criteria $1, \ldots, t$. The score of the rth alternative on kth criteria is named p_{rk}. We assume that a better score indicates a better performance. For the cases that a lower value represents a preferred outcome, such as diabetes A1c, we use $L - p_{rk}$ with large enough L so that all $L - p_{rk}$ are positive in our computation. We assume that among the criteria $C_1, C_2, \ldots, C_h$ the wins/losses are considered, for instance, in a ball game, only wins/losses matter. We use

$$q_k^{rs} = \begin{cases} 1, & \text{if } p_{rk} > p_{sk} \\ 0, & \text{if } p_{rk} = p_{sk} \\ -1, & \text{if } p_{rk} < p_{sk}. \end{cases}$$

We assume that for criteria $C_{h+1}, \ldots, C_m$ that the numerical difference scores are of interest. Let $\alpha_k = \max_r(p_{rk})$ and $\beta_k = \min_r(p_{rk})$, we define

$$p_k^{rs} = (1/2)\left(\frac{p_{rk} - p_{sk}}{\alpha_k - \beta_k} + 1\right).$$

Let w_k be a weight assigned on the kth criteria with $w_k \geq 0$ and $\sum_k w_k = 1$, then we defined

$$a_{r,s} = (1/2)\sum_{k=1}^{h} w_k(q_k^{rs}) + (1/2)\sum_{k=h+1}^{t} w_k(p_k^{rs}).$$

Then the matrix $A_{m \times m} = (a_{ij})$ can be used for ranking purpose. We assume that A is primitive. Let λ be the Perron root of A, and $v_0 = [1, 1, \ldots, 1]^T$. If $d = \lim_{n \to \infty}(A/\lambda)^n v_0 = [d_1, \ldots, d_m]^T$, where $d_r \geq 0$ for all r and at least one $d_r > 0$, then d is called a *ranking vector*. We have $P_r \geq P_s$ iff $d_r \geq d_s$.

Example 6.8: Let us consider the following table for knee replacement surgery data:

Hospital ID	Distance	Surgery quality	Patient rating	Price
A	80 miles	1.0	40%	$ 31,219
B	39 miles	0.7	100%	$ 27,179
C	49 miles	1.0	100%	$ 70,200
D	70 miles	1.0	80%	$ 18,766
E	51 miles	1.0	80%	$ 36,067
F	4 miles	1.0	80%	$ 37,392
G	62 miles	1.0	60%	$ 32,244
H	96 miles	1.0	100%	$ 20,826
I	70 miles	0.7	100%	$ 22,113
J	70 miles	0.7	100%	$ 26,009
K	51 miles	1.0	100%	$ 29,943
L	73 miles	0.7	80%	$ 24,870

Let R-vector be the Ranking vector, H-ID be the Hospital ID, and C_1 = distance, C_2 = surgery quality, C_3 = patient rating, C_4 = price. Let patient α's preference yield $w_1 = 0.08$, $w_2 = 0.42$, $w_3 = 0.25$, $w_4 = 0.25$. We have the following table of the top five selections for patient α,

Ranking	R-vector	H-ID	C_1	C_2	C_3	C_4
1	0.3499	H	96 miles	1.0	100%	$ 20,826
2	0.3486	K	51 miles	1.0	100%	$ 29,943
3	0.3351	D	70 miles	1.0	80%	$ 18,766
4	0.3254	F	4 miles	1.0	80%	$ 37,392
5	0.3153	E	51 miles	1.0	80%	$ 36,067

Let patient β's preference yield $w_1 = 0.1$, $w_2 = 0.3$, $w_3 = 0.15$, $w_4 = 0.5$. We have the following table of the top five selections for patient β,

Ranking	R-vector	H-ID	C_1	C_2	C_3	C_4
1	0.3436	D	70 miles	1.0	80%	\$ 18,766
2	0.3413	H	96 miles	1.0	100%	\$ 20,826
3	0.3297	K	51 miles	1.0	100%	\$ 29,943
4	0.3137	F	4 miles	1.0	80%	\$ 37,392
5	0.3026	E	51 miles	1.0	80%	\$ 36,067

The reader is referred to Ping Heidi Huang and T. T. Moh's article *A non-linear non-weight method for multi-criteria decision making*, Ann. Oper. Res. (2017) 248:239–251. ∎

Exercises

(1) Let A be the following matrix,

$$A = \begin{bmatrix} 1/2 & 1/2 \\ 1/2 & 1/2 \end{bmatrix}.$$

Find its **Perron root** and **Perron vector**.

(2) Let us use the notation of the preceding exercise. Let us follow our subsection on *Google rank* and *PageRank*. Show that the matrix $D = A$ and $D^m = A$ for all m. Let $\{w_1 = [1,1]^T, w_2 = [1,-1]^T\}$. Show that

$$\lim_{m \to \infty} D^m w_1 = w_1, \quad \lim_{m \to \infty} D^m w_2 = 0.$$

(3) Let A be the following matrix,

$$A = \begin{bmatrix} 0.4 & 0.3 & 0.2 & 0.1 \\ 0.3 & 0.2 & 0.1 & 0.4 \\ 0.2 & 0.1 & 0.4 & 0.3 \\ 0.1 & 0.4 & 0.3 & 0.2 \end{bmatrix}.$$

Find its **Perron root** and **Perron vector**.

(4) Let $P_1 = (3,7)$, $P_2 = (6,4)$, $P_3 = (2.9, 5.9)$, $P_4 = (1,9)$. Suppose that only wins/losses matter. Find their rank.

(5) In Example 6.8, let us consider a patient γ who treats all criteria as wins/losses only, and has a weight $w_1 = 0.01$, $w_2 = 0.4$, $w_3 = 0.09$, $w_4 = 0.5$. Find the top five alternatives.

(6) Find the **Perron root** and **Perron vector** of the following matrix

$$A = \begin{bmatrix} 4 & 3 & 3 \\ 3 & 2 & 3 \\ 2 & 4 & 3 \end{bmatrix}.$$

6.3 QR Method. Find the Eigenvalues

In the matrices theory, we have two kinds of problems: (1) Solve a system of linear equations. (2) Solve the eigen-problem,

$$(1) \ \mathbf{Ax = b} \quad \text{or} \quad (2) \ \mathbf{Bx = \lambda x}.$$

We will discuss the second problem in this section.

In the preceding section, we discussed the *power method*. For some matrix A, we have

$$\lim_{m \mapsto \infty} A^m v = w_1.$$

However, for general matrix A, the computations are complicated and the convergent rate is slow. During 1959–1961, J. G. F. Francis[5] of Ferranti Ltd., London, England discovered a practical QR method for computing eigenvalues, which is much faster. Note that this method is scientifically sound, although it has not been verified by pure mathematical reasoning.

The procedure of the QR method is as follows,

(1) Write $A_0 = A = QR = Q_0 R_0$, where Q_0 is unitary, and R_0 is upper-triangular. We may achieve this step by the Gram–Schmidt process.

(2) Define $A_1 = R_0 Q_0 = Q_1 R_1$ where Q_1 is unitary, and R_1 is upper-triangular. Note that $Q_0^{-1} A_0 Q_0 = A_1$, therefore A_1 is similar to A_0, and there is no change of eigenvalues.

---

[5]Francis, J. G. F. English computer scientist. 1934–.

(3) If we have $A_k = Q_k R_k$ defined, then we define $A_{k+1} = R_k Q_k$. Experimentally, it is known that in the complex cases, A_i approaches a triangular form, and in the real cases, A_i approaches a semi-triangular form (see the following definition).

Definition A real matrix A is in the semi-triangular form if

$$A = \begin{bmatrix} T_{11} & \cdots & \cdots & \cdots & T_{1n} \\ 0 & T_{22} & \cdots & \cdots & T_{2n} \\ \cdots & \cdots & \cdots & \cdots & \cdots \\ 0 & \cdots & 0 & \cdots & \cdots \\ 0 & 0 & \cdots & 0 & T_{nn} \end{bmatrix}$$

where T_{ii} is either a 1×1 or 2×2 matrix. ∎

Example 6.9: We may use the *Matlab* program to find the QR decomposition of some matrix. Let

$$A_0 = \begin{bmatrix} 1.0000 & 1.0000 & 0.0000 \\ -1.0000 & 1.0000 & 0.0000 \\ 5.0000 & 6.0000 & 2.0000 \end{bmatrix}.$$

It is not hard to see that the above 3×3 matrix has $1 \pm i, 2$ as the eigenvalues. We shall use the QR decomposition to find it without using the characteristic equation as follows. We first use *Matlab* to find Q_0 and R_0,

$$Q_0 = \begin{bmatrix} -0.1925 & 0.0514 & -0.9800 \\ 0.1925 & -0.9773 & -0.0891 \\ 0.9623 & 0.2057 & 0.1782 \end{bmatrix}$$

$$R_0 = \begin{bmatrix} -5.1962 & -5.7735 & -1.9245 \\ 0.0000 & -2.1602 & -0.4115 \\ 0.0000 & 0.0000 & 2.3563 \end{bmatrix}.$$

Thus we have

$$A_1 = R_0 Q_0 = \begin{bmatrix} 1.17407 & 5.7709 & 5.2635 \\ -0.0198 & 2.1958 & 0.1191 \\ -0.3429 & -0.0733 & 0.0635 \end{bmatrix}.$$

We may continue the above process and find

$$A_{10} = \begin{bmatrix} 2.5573 & 5.5500 & -5.2230 \\ -0.1537 & 0.4729 & 1.5010 \\ -0.0858 & -0.9640 & 0.9698 \end{bmatrix}$$

$$A_{20} = \begin{bmatrix} 2.0098 & 5.5209 & 5.5248 \\ -0.0537 & 0.9893 & 0.9893 \\ -0.0016 & -0.9990 & 1.0010 \end{bmatrix}$$

$$A_{30} = \begin{bmatrix} 1.9996 & -5.5226 & 5.5229 \\ -0.0001 & 1.0003 & 0.9997 \\ -0.0001 & -1.9990 & 1.0000 \end{bmatrix}$$

$$A_{40} = \begin{bmatrix} 2.0012 & 0.0186 & 7.8101 \\ -0.0000 & 1.0000 & 0.9988 \\ -0.0001 & -1.0000 & 0.9999 \end{bmatrix}$$

and we estimate that $\lambda_1 \approx 2.0012$, and λ_2, λ_3 are approximated by the roots of the following equation,

$$\det \begin{bmatrix} \lambda - 1.0000 & -0.9988 \\ 1.0000 & \lambda - 0.9999 \end{bmatrix}.$$

We can easily find $\lambda_2, \lambda_3 \approx 1 \pm i$. ∎

[The Shifted Algorithm]

If the number s_k is close to an eigenvalue, then we may shift the equations,

$$A_k = Q_k R_k + s_k I, \quad A_{k+1} = R_k Q_k + s_k I.$$

It is easy to see that

$$Q_k^{-1} A_k Q_k = A_{k+1}.$$

The experiments show that usually it speeds up the process of locating eigenvalues if we take s_k to be the (nn)th term of the matrix A_k, the so-called *Wilkinson shift*.[6] See the following example,

---

[6]Wilkinson, J. English numerical analyst. 1919–1986.

Example 6.10: We shall use the previous example. Let

$$A_0 = \begin{bmatrix} 1.0000 & 1.0000 & 0.0000 \\ -1.0000 & 1.0000 & 0.0000 \\ 5.0000 & 6.0000 & 2.0000 \end{bmatrix}$$

and $A_0 - 2I = Q_0 R_0$, $A_1 = R_0 Q_0 + 2I$, then we have

$$A_1 = \begin{bmatrix} -0.1111 & 5.3749 & 5.1949 \\ -1.4157 & 2.1111 & 2.1170 \\ 0 & 0 & 2.0000 \end{bmatrix}.$$

Just in one step, we find a characteristic root 2.   ■

Example 6.11: The QR method may not work for all square matrices. Let A be a unitary matrix. Then it is easy to see $A_0 = A = AI = IA = AI = \cdots$. It means that $A_0 = A_1 = \cdots$. Therefore the QR process will not produce any new matrix. If A is not upper-triangular to begin with, then the sequence $\{A_0, A_1, \ldots\}$ will not approximate any upper-triangular matrix.

Similarly for the Wilkinson shift. Let A be a unitary matrix with $a_{nn} = 0$. Then the shift is void, and we have the same conclusion.   ■

Householder Transformation

The question which bothers us is the size of the matrix A in computation. As the size increases, the number of terms will increase in the square of the size. To avoid the multiplication of non-zero terms, one way is to increase the number of zero terms. We want to select the coordinate system to increase the number of zero terms. That is the topic of this subsection. We give the following definitions about the shape of a matrix A.

Definition We have

(1) A matrix A is said to be *bi-diagonal* iff a_{ii} and $a_{i,i+1}$ (or $a_{i,i-1}$ exclusively) are possibly non-zero terms.

(2) A matrix A is said to be *tri-diagonal* iff a_{ii}, $a_{i,i+1}$ and $a_{i,i-1}$ are possibly non-zero terms.

(3) A matrix A is said to be *Hessenberg*[7] if for all $i \geq j + 2$, $a_{ij} = 0$. ∎

Given an $n \times n$ real or complex matrix A, we will find a coordinate system $\{e_i\}$ so that A is Hessenberg with respect to $\{e_i\}$. For this purpose we will define,

Definition A *Householder*[8] *transformation* is a transformation H_u of the following form,

$$H_u = I - 2\frac{uu^*}{u^*u}$$

where u is a non-zero vector in $\mathbb{C}^n$. ∎

We have the following proposition,

Proposition 6.3.1: *We have*

(1) $H_u(u) = -u$.
(2) $H_u(w) = w$, *for all* $\langle u|w \rangle = 0$.

Proof. (1) We have $H_u(u) = Iu - 2\frac{uu^*}{u^*u}u = u - 2\frac{u(u^*u)}{u^*u} = u - 2u = -u$. (2) Similar. ∎

Proposition 6.3.2: *The Householder transformation H_u is hermitian and unitary.*

Proof. It is easy to see that H_u is hermitian. We have

$$H_u^*H_u = (H_u)^2 = I - 4\frac{uu^*}{u^*u} + 4\frac{u(u^*u)u^*}{u^*uu^*u} = I.$$

Therefore H_u is unitary. ∎

In most textbooks, the following Proposition is proved only for the real case. We use a trick of D. F. Xu and J. Qian (*Six Lectures on Matrix Computations* (High Education Press, Beijing, China)) to prove the complex case which includes the real case.

---

[7]Hessenberg, K. German engineer. 1904–1959.
[8]Householder, A. American mathematician. 1904–1993.

Let $x = [x_1, \ldots, x_n]^T$ be any vector with norm $|x| = \sigma$ and $z = [1, 0, \ldots, 0]^T$. Let

$$\delta = \begin{cases} 1, & \text{if } x_1 = 0, \\ \overline{x_1}/|x_1|, & \text{if } x_1 \neq 0. \end{cases}$$

Note that $|\delta| = 1$ i.e., $\delta\bar{\delta} = 1$.

Let $u = \delta x + \sigma z$. Note that we always have

$$(\delta x + \sigma z)^*(\delta x + \sigma z) = 2\sigma(\sigma + |x_1|).$$

For a proof, note that if $x_1 = 0$, then $\delta = 1$ and $z^*x = x_1 = 0 = \bar{x}_1 = x^*z$. Our formula is obvious. If $x_1 \neq 0$, then $\delta = \overline{x_1}/|x_1|$ and $(\sigma z)^* \delta x = \sigma \delta x_1 = \sigma|x_1| = (\delta x)^* \sigma z$. Our formula follows.

Then we have

Proposition 6.3.3: *We always have $H_u(x) = -\bar{\delta}\sigma z$.*

Proof. We have the following computations

$$H_u(x) = Ix - 2(\delta x + \sigma z)\frac{(\delta x + \sigma z)^* x}{(\delta x + \sigma z)^*(\delta x + \sigma z)}$$

$$= x - (\delta x + \sigma z)\frac{(\delta x + \sigma z)^* x}{\sigma^2 + \sigma|x_1|}$$

$$= x - (\delta x + \sigma z)\frac{\bar{\delta}x^*x + \sigma x_1}{\sigma^2 + \sigma|x_1|}$$

$$= x - (\delta x + \sigma z)\frac{\bar{\delta}\sigma^2 + \sigma x_1}{\sigma^2 + \sigma|x_1|}$$

$$= x - (\delta x + \sigma z)\frac{\bar{\delta}\sigma + x_1}{\sigma + |x_1|}$$

$$= x - (\delta x + \sigma z)\bar{\delta}$$

$$= x - (x + \bar{\delta}\sigma z)$$

$$= -(\bar{\delta}\sigma z). \qquad \blacksquare$$

The important proposition is the following proposition which shows that any square matrix A is similar to a Hessenberg matrix.

Proposition 6.3.4: *Let $A_{n \times n}$ be a complex square matrix. Then there is a unitary matrix U such that U^*AU is Hessenberg.*

Proof. Let $A = (a_{ij}) = (c_1, \ldots, c_n)$. Let $x = [a_{21}, a_{31}, \ldots, a_{n1}]^T$ be a vector in $\mathbb{C}^{n-1}$. Note that x is c_1 omitting the first entry a_{11}. Let z, σ, u be defined as in the preceding paragraphs. Let us define

$$
U_1 = \begin{bmatrix} 1 & 0 & \cdots & 0 & 0 \\ 0 & & & & \\ 0 & & & & \\ \cdot & & H_u & & \\ 0 & & & & \\ 0 & & & & \end{bmatrix}.
$$

Then we have $A_1 = U_1 A U_1 = [U_1 c_1, \ldots, U_1 c_n] U_1$. While $U_1 c_1 = [a_{11}, -\bar{\delta}\sigma, 0, \ldots, 0]^T$. It is easy to see that

$$
A_1 = \begin{bmatrix} a_{11} & * & \cdots & * & * \\ -\bar{\delta}\sigma & & & & \\ 0 & & & & \\ \cdot & & B_1 & & \\ 0 & & & & \\ 0 & & & & \end{bmatrix}.
$$

Now we shall work on the smaller matrix B_1. We shall find U_2 of the following form

$$
U_2 = \begin{bmatrix} 1 & 0 & \cdots & 0 & 0 \\ 0 & 1 & 0 & \cdots & 0 \\ 0 & 0 & & & \\ \cdot & & & H_2 & \\ 0 & 0 & & & \\ 0 & 0 & & & \end{bmatrix}.
$$

We simply let $U = \prod H_{u_i}$, and UAU is the required Hessenberg matrix. $\blacksquare$

If A is hermitian, then $UAU = U^*AU$ stays hermitian, and will be tri-diagonal. The number of non-zero terms will be reduced from n^2 to $3n - 2$ — that is a huge reduction of computations. In the computations of matrix A, we shall replace it by the Hessenberg form of it. The important property is that in the QR decomposition, the matrix Q will remain Hessenberg. We have

Lemma 6.3.5: *If R is an invertible upper-triangular matrix, then R^{-1} is upper-triangular.*

Proof. Left to the reader as an exercise.   ∎

Lemma 6.3.6: *If A is Hessenberg, R^{-1} is upper-triangular. Then $Q = AR^{-1}$ is Hessenberg.*

Proof. Left to the reader as an exercise.   ∎

Lemma 6.3.7: *If Q is Hessenberg, R is upper-triangular. Then RQ is Hessenberg.*

Proof. Left to the reader as an exercise.   ∎

Repeatedly applying the above lemmas, we conclude that,

Proposition 6.3.8: *Let A be a real or complex non-singular square Hessenberg matrix. Then in the QR series, and in the Shifted QR series, we have all A_i, Q_i being Hessenberg.*

Proof. See the above.   ∎

Remark: We have already showed that any polynomial $f(x) = x^n + a_1 x^{n-1} + \cdots + a_n$ is the characteristic polynomial $\chi_A(\lambda)$ of the following associated matrix A (cf. Example in Section 3.4),

$$
A = \begin{bmatrix}
0 & 0 & \cdot & \cdot & 0 & -a_m \\
1 & 0 & \cdot & \cdot & \cdot & -a_{m-1} \\
0 & 1 & \cdot & \cdot & \cdot & -a_{m-2} \\
\cdot & & \cdot & \cdot & & \cdot \\
0 & 0 & \cdot & \cdot & 1 & -a_1
\end{bmatrix}.
$$

To solve a polynomial $f(x) = x^n + a_1 x^{n-1} + \cdots + a_n = 0$, the software *Matlab* first locates its associated matrix A as above, then it finds the eigenvalues λ_i of the associated matrix A by the above-mentioned QR method. Certainly it pays attention to the special form of A as

$$A = \begin{bmatrix} 0 & 0 & \cdot & \cdot & 0 & 1 \\ 1 & 0 & \cdot & \cdot & \cdot & 0 \\ 0 & 1 & \cdot & \cdot & \cdot & 0 \\ \cdot & \cdot & \cdot & \cdot & \cdot & \cdot \\ 0 & 0 & \cdot & \cdot & 1 & 0_1 \end{bmatrix} - \begin{bmatrix} a_n + 1 \\ a_{n-1} \\ a_{n-2} \\ \cdot \\ a_1 \end{bmatrix} \begin{bmatrix} 0, 0, \ldots, 0, 1 \end{bmatrix}. \qquad \blacksquare$$

[Krylov Subspace]

In 1931, Nicolai Krylov[9] found a way to compute the characteristic polynomial $\chi_A(\lambda)$, given an $n \times n$ matrix A. His method is for a general vector v, where the sequence of vectors $\{v, Av, \ldots, A^{n-1}v\}$ is likely to be linearly independent. Therefore $A^n v$ has an expression as

$$A^n v + a_1 A^{n-1} v + \cdots + a_{n-1} Av + a_n v = 0.$$

It is easy to prove that

$$\chi_A(\lambda) = \pm(\lambda^n + a_1 \lambda^{n-1} + \cdots + a_{n-1}\lambda + a_n).$$

In 1950, Magnus Hestenes,[10] Edward Stiefel[11] and Cornelius Lanczos[12] (once at Purdue University), all from the Institute for Numerical Analysis at the National Bureau of Standards, USA, initiated the development of the Krylov subspace iteration methods to approximate the eigenvalues of a large ($n = 10^6$ or more) and sparse matrix A. A matrix A is said to be *sparse* if almost 10% of its entries are non-zeroes. Usually it takes years for the fastest computer to carry out the QR procedure, hence it is impractical.

Pick up a vector v, consider the smaller and computable subspace $V_k = \{v, Av, \ldots, A^{k-1}v\}$. We want to project the matrix A to the

---

[9]Krylov, N. Soviet mathematician. 1941–.

[10]Hestenes, M. American mathematician. 1906–1992.

[11]Stiefel, E. Swiss mathematician. He is known for the Stiefel manifold in pure mathematics. 1909–1978.

[12]Lanczos, C. Jewish–Hungarian mathematician and physicist. 1893–1974.

smaller subspace. Let $\{q_1, q_2, \ldots, q_m\}$ be an orthonormal basis for V_k, we pick up the matrix $Q = [q_1, q_2, \ldots, q_m]$. Then the matrix Q^*AQ is the projection of A to V_k which is an $m \times m$ matrix. Suppose that the number m is small enough such that we may use the QR method to find its eigenvalues and eigenvectors. We will ponder the following,

Questions: The selection of vector v, the integer k and the matrix Q such that some eigenvalues and eigenvectors of Q^*AQ will approximate the eigenvalues and the eigenvectors we need for the original problem of A. Furthermore, what is the quantitative approximation?

The discussions of the above are beyond the scope of the present book.

Exercises

(1) Let $x \neq y$ be two vectors in a finite dimensional inner product space V and $|x| = |y|$. Let $v = x - y$. Show that the corresponding Householder transformation H will map x to y and y to x.

(2) Prove that if R is an invertible upper-triangular matrix, then R^{-1} is upper-triangular.

(3) Prove that if A is Hessenberg, R^{-1} is upper-triangular. Then $Q = AR^{-1}$ is Hessenberg.

(4) Prove that if Q is Hessenberg, R is upper-triangular. Then RQ is Hessenberg.

(5) Let A be the following 5×5 matrix,

$$A = \begin{bmatrix} 1 & 1 & 1 & 1 & 1 \\ 1 & 0 & 0 & 0 & 0 \\ 1 & 0 & 0 & 0 & 0 \\ 1 & 0 & 0 & 0 & 0 \\ 1 & 0 & 0 & 0 & 0 \end{bmatrix}.$$

Show that there is an orthogonal matrix Q such that $Q^{-1}AQ$ is tri-diagonal.

6.4 Approximate Solutions to a System of Linear Equations

In this section we will handle the problem of solving a system of linear equations $\mathbf{Ax = b}$. Theoretically, we have the *Gauss–Jördan* process to solve a system of linear equations. The time of computations is approximately $n^3/3$, where n is the number of variables. For small-sized questions, say n is less than a few thousands, the standard method works fine. However if $n = 10^7$, then we need $10^{21}/3$ computations. If we use a fast computer which performs 10^{12} computations per second, it requires 30 years of computations, hence it is impractical! For a large scale system of linear equations, we require it to be sparse, say there are at most 10% non-zero coefficients, we will use iterative methods, which either produce an exact solution or an approximate solution fast.

We shall first discuss three iteration methods.

Jacobi[13] Method

In some situations, say $A = (a_{ij})$ is diagonal dominated, i.e., $|a_{ii}| > \sum_{i \neq j} |a_{ij}|$ for all i, we may write $A = D + R$, and we require all diagonals of A to be non-zero, where D is the diagonal part of A, and R is the remaining part of A. Let us assign x_0 to be any initial vector, and define

$$Dx^{(k+1)} = b - Rx^{(k)}.$$

If $\lim_{k \to \infty} x^{(k)} = x$, then we have $Dx = b - Rx$ or $Ax = Dx + Rx = b$, and x is a solution of $Ax = b$. The above equation may further be written as

$$x^{(k+1)} = D^{-1}b - D^{-1}Rx^{(k)}.$$

Or analytically it is

$$x_i^{(k+1)} = \frac{1}{a_{ii}} \left(b_i - \sum_{i \neq j} a_{ij} x_j^{(k)} \right), \quad i = 1, 2, \ldots, n.$$

---

[13] Jacobi, C. German mathematician. 1804–1851.

Example 6.12: Let us consider a simple example.

$$A = \begin{bmatrix} 3 & 1 \\ 2 & 5 \end{bmatrix}, \quad b = \begin{bmatrix} 2 \\ 3 \end{bmatrix}, \quad \text{and} \quad x^{(0)} = \begin{bmatrix} 1 \\ 1 \end{bmatrix},$$

then we have

$$D^{-1} = \begin{bmatrix} 1/3 & 0 \\ 0 & 1/5 \end{bmatrix}, \quad L = \begin{bmatrix} 0 & 0 \\ 2 & 0 \end{bmatrix}, \quad \text{and} \quad U = \begin{bmatrix} 0 & 1 \\ 0 & 0 \end{bmatrix},$$

and $D^{-1}b - D^{-1}Rx^{(k)}$ can be written as

$$D^{-1}b = \begin{bmatrix} 1/3 & 0 \\ 0 & 1/5 \end{bmatrix} \begin{bmatrix} 2 \\ 3 \end{bmatrix} = \begin{bmatrix} 0.6666 \\ 0.6 \end{bmatrix}.$$

We apply the formula $x^{(k+1)} = D^{-1}(b - Rx^{(k)})$ and get

$$x^{(1)} = \begin{bmatrix} 0.3333 \\ 0.2 \end{bmatrix},$$

and

$$x^{(10)} = \begin{bmatrix} 0.5384 \\ 0.3847 \end{bmatrix},$$

which is very close to the true solution

$$x = \begin{bmatrix} 0.5385 \\ 0.3846 \end{bmatrix}. \qquad \blacksquare$$

Gauss[14]–Seidel[15] Method

At least we require that a_{ii} be non-zero for all i. We rewrite $A = L_* + R$ where L_* is the lower-triangular part of A and R is the strict upper-triangular part of A. Then we pick up an initial vector $v^{(0)}$ and define

$$L_* x^{(k+1)} = b - Rx^{(k)}.$$

If $\lim_{k \mapsto \infty} x^{(k)} = x$, then we have $L_* x = b - Rx$ or $Ax = L_* x + Rx = b$, and x is a solution of $Ax = b$. The above equation may

---

[14]It is mentioned in a private letter from Gauss to his student Gerling in 1823.
[15]Seidel, P. German mathematician. He published it in 1874. 1821–1896.

further be written as

$$x^{(k+1)} = L_*^{-1}b - L_*^{-1}Rx^{(k)}.$$

Or analytically it is

$$x_i^{(k+1)} = \frac{1}{a_{ii}}\left(b_i - \sum_{j=1}^{i-1} x^{(k+1)} - \sum_{j=i+1}^{n} a_{ij}x_j^{(k)}\right), \quad i = 1, 2, \ldots, n.$$

As clearly indicated by the above analytical formula, when we compute $x_{i+1}^{(k+1)}$, we may discard $x_{i+1}^{(k)}$ and retain $x_i^{(k+1)}$. Thus we will save some storage space.

Successive Over-relaxation

This method is similar to the two above methods. We write the matrix as $A = D + L + U$ where L is the strict lower-triangular part of A. Let the *relaxation factor* $w > 1$ be selected. Then we have

$$(D + wL)x = wb - (wU + (w - 1)D)x.$$

As usual, we will consider the following iterative equation

$$(D + wL)x^{(k+1)} = wb - (wU + (w - 1)D)x^{(k)}.$$

If $\lim_{k \mapsto \infty} x^{(k)} = x$, then we have $wAx = w(Dx + Lx + Ux) = wb$, and x is a solution of $Ax = b$. The above equation may further be written as

$$x^{(k+1)} = (D + wL)^{-1}(wb - (wU + (w - 1)D))\,x^{(k)}.$$

Or analytically it is

$$x_i^{(k+1)} = (1 - w)x_i^{(k)} + \frac{w}{a_{ii}}\left(b_i - \sum_{j<i} a_{ij}x_j^{(k+1)} - \sum_{j>i} a_{ij}x_j^{(k)}\right),$$
$$i = 1, 2, \ldots, n.$$

[Conjugate Gradient Method]

This method was developed for solving an equation of the form $Ax = b$ with A a real symmetric and positive-definite matrix by Magnus Hestenes, Edward Stiefel and Cornelius Lanczos. It happens many times in the real world of physics and engineering that the equation is of the form $Ax = b$ with A a real symmetric and positive-definite matrix and of large size (say, bigger than 10^6). For instance, a finite difference problem can be represented by $Ax = b$, where

$$A = \begin{bmatrix} 2 & -1 & 0 & \cdots & \cdots & \cdots & 0 \\ -1 & 2 & -1 & 0 & \cdots & \cdots & 0 \\ 0 & -1 & 2 & -1 & 0 & \cdots & 0 \\ \cdots & \cdots & \cdots & \cdots & \cdots & \cdots & \cdots \\ \cdots & \cdots & \cdots & \cdots & \cdots & \cdots & \cdots \\ 0 & \cdots & \cdots & 0 & -1 & 2 & -1 \\ 0 & \cdots & \cdots & \cdots & 0 & -1 & 2 \end{bmatrix}$$

which is symmetric and positive-definite. We further note that to solve $Ax = b$ is equivalent to finding the minimal point of $f = \frac{1}{2}x^T A x - x^T b$.

The equation $Ax = b$ has a solution point in $\mathbb{R}^n$, while $\mathbb{R}^n$ is homogeneous and isotropic, therefore it is not easy to find a sequence of approximate points. On the other hand, geometrically, on the non-homogeneous and non-isotropic manifold (x, f) with a unique minimal point, starting with any point $(x_0, f(x_0))$, it is possible to slide down the quadratic hypersurface defined by $(x, f(x))$ by successively studying the *gradient vectors* at the points to locate the minimal point of (x, f).

If the size of A is too big to solve $Ax = b$ directly, we may find the solutions of $Ax = b$ by finding a sequence of approximate points to the minimal point of (x, f). This method is meaningful only for sparse matrix for $n = 10^6$ or larger.

Since A is real symmetric and positive-definite, the matrix A naturally defines an *inner product*, hence a geometry in $\mathbb{R}^n$. We have then two definitions of lengths. We have two different sense of angles. We will say that two vectors v, u are *orthogonal* if $v^T u = 0$, we will say that two vectors v, u are *conjugate* if $\langle v, u \rangle_A = v^T A u = 0$. The term *conjugate* means A-orthogonal. We hope that it explains the term *Conjugate Gradient Method*.

We know from Section 4.6 that $\text{grad}(f(x)) = Ax - b$. Start from any arbitrary initial point x_0 (we may take $x_0 = 0$), the negative of the gradient direction is $-(Ax_0-b)$. If it is 0, then we find the solution of the equation. Assume that $-(Ax-b) \neq 0$. Call the direction of $b - Ax_0 = r_0(= d_0)$ the *residual direction* which is the steepest downward direction. Now we shall move along the direction d_0 for a distance $\alpha_0 d_0$. The question is how long we should move? What is α_0?

Let us take the direction $d_0 = b - Ax_0(= r_0)$ (opposite to the gradient direction), and call r_0 the *residual*. Let us define *error* $e_0 = x_0 - x$ with x the true solution (which is unknown). Note that $r_0 = -Ae_0$, we think that r_0 is the image of e_0 under A. Moreover, $d_0 = r_0 = -f'(x_0) = -(-b + Ax_0)$, d_0 is the direction of the steepest descent.

We make a ray along the direction of d_0. We want to reach the lowest point along the ray. Let

$$x_1 = x_0 + \alpha_0 d_0, \quad r_1 = b - Ax_1.$$

The α_0 minimizes f when the directional derivative $\frac{d}{d\alpha_0} f(x_1)$ is equal to 0, which is $\frac{d}{d\alpha_0} f(x_1) = f'(x_1)^T \frac{d}{d\alpha_0} x_1 = f'(x_1)^T d_0 = -r_1^T d_0$. This property decides α_0 as follows, with all equivalent equalities,

$$r_1^T d_0 = 0$$
$$(b - Ax_1)^T d_0 = 0$$
$$(b - A(x_0 + \alpha_0 d_0)^T d_0 = 0$$
$$(b - Ax_0)^T r_0 - \alpha_0 (Ad_0)^T d_0 = 0$$
$$(b - Ax_0)^T r_0 = \alpha_0 (Ad_0)^T d_0$$
$$r_0^T r_0 = \alpha_0 d_0^T (Ad_0)$$

$$\alpha_0 = \frac{r_0^T r_0}{d_0^T Ad_0} = \frac{\langle r_0, r_0 \rangle}{\langle d_0, d_0 \rangle_A}.$$

[Inductive Definition of Conjugate Gradient Method]

Note that in the above computation, we pick up arbitrary x_0, then we have $d_0 = r_0(= b - Ax_0)$ and α_0 defined as above. While the term β_0 is not needed and hence undefined. In general we have for $j = 0, \ldots, i$ with x_j the jth approximation, we have $-f'(x_j) = b - Ax_j = r_j$ and the direction d_j. Suppose we have x_k, r_k, β_k, d_k for $k = 1, \ldots, i$ and α_s for $s = 0, \ldots, i-1$ defined. We will define the next set of numbers as,

$$x_0 \text{ arbitrary}, \quad d_0 = r_0 = b - Ax_0 \tag{1}$$

$$\alpha_i = \frac{\langle r_i, r_i \rangle}{\langle d_i, d_i \rangle_A} = \frac{r_i^T r_i}{d_{i-1}^T A d_{i-1}} \tag{2}$$

$$x_{i+1} = x_i + \alpha_i d_i \tag{3}$$

$$r_{i+1} = b - Ax_{i+1} = r_i - \alpha_i A d_i \tag{4}$$

$$\beta_{i+1} = \frac{r_{i+1}^T r_{i+1}}{r_i^T r_i} = \frac{\langle r_{i+1}, r_{i+1} \rangle}{\langle r_i, r_i \rangle} \tag{5}$$

$$d_{i+1} = r_{i+1} + \beta_{i+1} d_i. \tag{6}$$

We shall show that the definition of *Conjugate Gradient Method* is consistent. Note that if $r_k \neq 0$ for all $k < i$ and $r_i = 0$, then x_i is an exact solution of the equation $Ax - b = 0$ and we are done. Therefore we shall assume that $r_i \neq 0$, β_{i+1} is defined. We **claim** the denominator of α_i is not zero. We shall consider the problem if α_i is defined, or if $d_i A d_i$ is zero, i.e., $|d_i|_A$ is zero, iff d_i is zero. The next lemma will finish our **claim**. Inductively, we assume that the definition of *Conjugate Gradient Method* is consistent for all x_j, r_j, β_j, d_j for $j \leq i$, i.e., they are all defined for $j \leq i$, furthermore α_j is defined for all $j < i$ and $r_i \neq 0$. Note that β_{i+1} is already defined. Now we shall handle the next stage i.e., we want to handle α_i next. Then $x_{i+1}, r_{i+1}, d_{i+1}$ follows. We then finish the next stage. We have the following inductive lemmas with the above Eqs. (1)–(6) true for all indices $j < i$.

Lemma 6.4.1: *If $r_j \neq 0$ for all $j < i$, then we have*

(1) $d_{i-1}^T r_i = 0$.

(2) $d_i^T r_i = r_i^T d_i - r_i^T r_i$.

(3) $r_i = 0 \Leftrightarrow d_i = 0$.

Proof. (1) We shall use induction and assume that $d_{i-2}^T r_{i-1} = 0$, which naturally means $r_{i-1}^T d_{i-2} = 0$. We have

$$(d_{i-1}^T r_i)^T = r_i^T d_{i-1} = r_{i-1}^T d_{i-1} - \alpha_{i-1} d_{i-1}^T A d_{i-1}$$

$$= r_{i-1}^T d_{i-1} - r_{i-1}^T r_{i-1} = r_{i-1}^T (d_{i-1} - r_{i-1})$$

$$= \beta_{i-1} r_{i-1}^T d_{i-2} = 0.$$

(2) We have $d_i^T r_i = (r_i + \beta_i d_{i-1})^T r_i = r_i^T r_i$.

(3) ($\Rightarrow$) We have $r_i = 0 \Rightarrow \beta_i = 0 \Rightarrow d_i = r_i + \beta_i d_{i-1} = 0$.

($\Leftarrow$) It follows from (2) that $d_i = 0 \Rightarrow |r_i|^2 = 0 \Rightarrow r_i = 0$. ∎

We conclude that α_i is defined, it follows that $x_{i+1}, r_{i+1}, \beta_{i+1}, d_{i+1}$ are defined. Now we know that the *Conjugate Gradient Method* makes sense.

[*A*-orthogonal]

Does it go on forever? We want to show that $\{d_0, d_1, \ldots, d_i\}$ is a set of *A*-orthogonal vectors if all are not zeroes. Since our space is n-dimensional, then there are at most n *A*-orthogonal vectors. Thus if $d_j \neq 0$ for $j < n$, then d_n must be zero, hence x_n must be the exact solution. Therefore the *Conjugate Gradient Method* naturally stops. We first prove some lemmas.

Lemma 6.4.2: If $d_k^T A d_l = 0$ and $r_k \neq 0$ for all $k \neq l < i$, then we have $r_{j-1}^T A d_{j-1} = d_{j-1}^T A d_{j-1}$ for all $j \leq i$.

Proof. We have $r_{j-1}^T A d_{j-1} = (d_{j-1} - \beta_{j-1} d_{j-2})^T A d_{j-1} = d_{j-1}^T A d_{j-1}$. ∎

Lemma 6.4.3: We have $\beta_i = -1 + \dfrac{\alpha_{i-1}^2 d_{i-1}^T A^2 d_{i-1}}{r_{i-1}^T r_{i-1}}$.

Proof. Since it follows from Eq. (2) that $\alpha_{i-1} d_{i-1}^T A d_{i-1} = r_{i-1}^T r_{i-1}$. Thus it follows from Lemma 6.4.2 that $\alpha_{i-1} r_{i-1}^T A d_{i-1} = r_{i-1}^T r_{i-1}$. We have

$$\begin{aligned}
\beta_i &= \frac{r_i^T r_i}{r_{i-1}^T r_{i-1}} \\[2mm]
&= \frac{(r_{i-1} - \alpha_{i-1} A d_{i-1})^T (r_{i-1} - \alpha_{i-1} A d_{i-1})}{r_{i-1}^T r_{i-1}} \\[2mm]
&= \frac{r_{i-1}^T r_{i-1} - 2\alpha_{i-1} r_{i-1}^T A d_{i-1} + \alpha_{i-1}^2 d_{i-1}^T A^2 d_{i-1}}{r_{i-1}^T r_{i-1}} \\[2mm]
&= -1 + \frac{\alpha_{i-1}^2 d_{i-1}^T A^2 d_{i-1}}{r_{i-1}^T r_{i-1}}.
\end{aligned}$$

∎

Lemma 6.4.4: *We have:* (1) $d_i = r_i + \beta_i d_{i-1} = r_{i-1} - \alpha_{i-1} A d_{i-1} + \beta_i d_{i-1} = d_{i-1} - \beta_{i-1} d_{i-2} - \alpha_{i-1} A d_{i-1} + \beta_i d_{i-1}$. (2) $\alpha_{i-1} A d_{i-1} = d_{i-1} - \beta_{i-1} d_{i-2} + \beta_i d_{i-1} - d_i$. *Or* $\alpha_{i-1} d_{i-1}^T A = d_{i-1}^T - \beta_{i-1} d_{i-2}^T + \beta_i d_{i-1}^T - d_i^T$. ∎

Lemma 6.4.5: *We have* $\alpha_{i-2} d_{i-2}^T A = d_{i-2}^T - \beta_{i-2} d_{i-3}^T + \beta_{i-1} d_{i-2}^T - d_{i-1}^T$.

Proof. It follows from the preceding lemma. ∎

Lemma 6.4.6: *We have* $\frac{\alpha_{i-1}}{\alpha_{i-2}} d_{i-1}^T A d_{i-1} = \beta_{i-1} d_{i-2}^T A d_{i-2}$.

Proof. Since $r_j \neq 0$ and $d_j \neq 0$ (cf. Lemma 6.4.1) and $d_j^T A d_j \neq 0$ for all $j < i$, then we have

$$\frac{\frac{r_{i-1}^T r_{i-1}}{d_{i-1}^T A d_{i-1}}}{\frac{r_{i-2}^T r_{i-2}}{d_{i-2}^T A d_{i-2}}} d_{i-1}^T A d_{i-1} = \beta_{i-1} d_{i-2}^T A d_{i-2}.$$ ∎

We have the following proposition,

Proposition 6.4.7: *If* $r_k \neq 0$ *and* $d_k^T A d_l = 0$ *for all* $k \neq l < i$, *then we have* $d_{i-1}^T A d_i = 0$.

Proof. We make an induction on i. We have
$$d_{i-1}^T A d_i = d_{i-1}^T A (r_{i-1} - \alpha_{i-1} A d_{i-1} + \beta_i d_{i-1})$$
$$= d_{i-1}^T A d_{i-1} - \alpha_{i-1} d_{i-1}^T A^2 d_{i-1} + \beta_i d_{i-1}^T A d_{i-1}$$
$$= d_{i-1}^T A d_{i-1} - \alpha_{i-1} d_{i-1}^T A^2 d_{i-1} - 1$$
$$+ \frac{\alpha_{i-1}^2}{r_{i-1}^T r_{i-1}} d_{i-1}^T A^{2d_{i-1}} d_{i-1}^T A d_{i-1}$$
$$= 0.$$ ∎

Proposition 6.4.8: *If* $r_k \neq 0$ *and* $d_k^T A d_l = 0$ *for all* $k \neq l < i$, *then we have* $d_{i-2}^T A d_i = 0$.

Proof. We make an induction on i. We have

$$d_{i-2}^T A d_i = d_{i-2}^T A (d_{i-1} - \alpha_{i-1} A d_{i-1} + \beta_i d_{i-1} - \beta_{i-1} d_{i-2})$$

$$= d_{i-2}^T A d_{i-1} - \alpha_{i-1} d_{i-2}^T A^2 d_{i-1}$$

$$+ \beta_i d_{i-2}^T A d_{i-1} - \left(-1 + \frac{\alpha_{i-2}^2 d_{i-2}^T A^2 d_{i-2}}{r_{i-2}^T r_{i-2}} \right) d_{i-2}^T A d_{i-2}$$

$$= -\alpha_{i-1} d_{i-2}^T A^2 d_{i-1} - \left(-1 + \frac{\alpha_{i-2}^2 d_{i-2}^T A^2 d_{i-2}}{r_{i-2}^T r_{i-2}} \right) d_{i-2}^T A d_{i-2}$$

$$= -\alpha_{i-1} d_{i-2}^T A^2 d_{i-1} + d_{i-2}^T A d_{i-2} - \alpha_{i-2} d_{i-2}^T A^2 d_{i-2}$$

$$= \frac{\alpha_{i-1}}{\alpha_{i-2}} d_{i-1}^T A d_{i-1} + d_{i-2}^T A d_{i-2}$$

$$- d_{i-2}^T A d_{i-2} - \beta_{i-1} d_{i-2}^T A d_{i-2}$$

$$= 0. \qquad \blacksquare$$

Proposition 6.4.9: *If $r_k \neq 0$ and $d_k^T A d_l = 0$ for all $k \neq l < i$, then we have $d_j^T A d_i = 0$ for all $i - j \geq 3$.*

Proof. We make an induction on i. We have

$$d_j^T A d_i = d_j^T A (d_{i-1} - \alpha_{i-1} A d_{i-1} + \beta_i d_{i-1} - \beta_{i-1} d_{i-2})$$

$$= -\alpha_{i-1} d_j^T A A d_{i-1}$$

$$- \alpha_{i-1} (d_j^T - \beta_j d_{j-1}^T + \beta_{j+1} d_j^T - d_{j+1}^T) A d_{i-1}$$

$$= 0. \qquad \blacksquare$$

Theorem 6.4.10: *If $r_k \neq 0$ for all $k \neq l < i$, then we have $d_j^T A d_s = 0$ for all $s \neq j \leq i$.* $\qquad \blacksquare$

Hence the above process must terminate after at most n steps. This is true mathematically, however due to round-off errors, it can take more than n steps (or fail to converge) in computations.

[Krylov Subspace]

We have the following proposition,

Proposition 6.4.11: *If $x_j \neq 0$ for $j = 0, 1, \ldots, i+1$, then the two vector subspaces $\langle r_0, r_1, \ldots, r_{i+1} \rangle = \langle d_0, d_1, \ldots, d_{i+1} \rangle (= V_{i+1})$.*

Proof. It follows from Eq. (6).   ∎

Proposition 6.4.12: *Let us use the notations of Proposition 6.4.11. We have*

$$V_{i+1} = \langle d_0, A d_0, A^2 d_0, \ldots, A^i d_0 \rangle$$
$$= \langle r_0, A r_0, A^2 r_0, \ldots, A^i r_0 \rangle.$$

Furthermore if $x_0 = 0$, *then we have* $r_0 = b$, *and* $V_{i+1} = \langle b, Ab, A^2 b, \ldots, A^i b \rangle$.

Proof. It is left to the reader.   ∎

Therefore we may think that the *Conjugate Gradient Method* live in some Krylov subspace. We have the following definition,

Definition We define the *Krylov sequence* $x^{(1)}, x^{(2)}, \ldots,$ as

$$x^{(k)} = \operatorname*{argmin}_{x \in V_k} f(x) = \operatorname*{argmin}_{x \in V_k} \|x - x^*\|_A^2$$

where $\operatorname{argmin}_{x \in V_k} f(x)$ means $\{x : f(x)$ is minimal for all $x \in V_k\}$ as usual.   ∎

Then we have

- $f(x^{(k+1)}) \le f(x^{(k)})$.
- $x^{(n)} = x^*$.
- $x^{(k)} = p_k(A)b$, where p_k is a polynomial of degree $< k$.

Exercises

(1) Given that the following $n \times n$ matrix A,

$$A = \begin{bmatrix} 2 & -1 & 0 & \cdots & \cdots & \cdots & 0 \\ -1 & 2 & -1 & 0 & \cdots & \cdots & 0 \\ 0 & -1 & 2 & -1 & 0 & \cdots & 0 \\ \cdots & \cdots & \cdots & \cdots & \cdots & \cdots & \cdots \\ \cdots & \cdots & \cdots & \cdots & \cdots & \cdots & \cdots \\ 0 & \cdots & \cdots & 0 & -1 & 2 & -1 \\ 0 & \cdots & \cdots & \cdots & 0 & -1 & 2 \end{bmatrix}.$$

Let the Jacobi matrix $J = D^{-1}(-R)$ where D is the diagonal part of A and $R = A - D$. Find J and show that $v_1 = [\sin(\pi h),$

$\sin(2\pi h), \ldots, \sin(n\pi h)]^T$ is an eigenvector of J associated with eigenvalue $\lambda_1 = \cos(\pi h)$ where $h = 1/(n+1)$.

(2) Show that the following $n \times n$ matrix A is symmetric and positive-definite for any n,

$$A = \begin{bmatrix} 2 & 1 & 0 & \cdots & \cdots & \cdots & 0 \\ 1 & 2 & 1 & 0 & \cdots & \cdots & 0 \\ 0 & 1 & 2 & 1 & 0 & \cdots & 0 \\ \cdots & \cdots & \cdots & \cdots & \cdots & \cdots & \cdots \\ \cdots & \cdots & \cdots & \cdots & \cdots & \cdots & \cdots \\ 0 & \cdots & \cdots & 0 & 1 & 2 & 1 \\ 0 & \cdots & \cdots & \cdots & 0 & 1 & 2 \end{bmatrix}.$$

(3) If you have a computer which does 10^{12} multiplications a second and you want to solve the equation $Ax = b$ where A is the matrix in (1) and with $n = 10^7$. If you use the *Conjugate Gradient Method*, and compute to n steps. How long do you have to compute?

(4) Finish the proof of Proposition 6.4.4.

(5) Finish the proof of Proposition 6.4.12.

(6) Finish the proof of Proposition 6.4.13.

(7) Finish the proof of Proposition 6.4.14.

(8) Prove that $f(x^{(k+1)}) \leq f(x^{(k)})$ in the Krylov space.

6.5 Linear Programming

In this section, we only consider the real field $\mathbb{R}$. In mathematics and the natural sciences, there are many max–min problems such as the refraction of light and max–min problems about eigenvalues in Section 4.8. In the real world, there are many max–min problems involving linear inequalities. It started when Fourier[16] published a work in 1827 on solving a system of linear inequalities. In 1939, Soviet economist L. Kantorovich (1912–1986) formulated a problem during World War II, to plan expenditures in order to reduce the

---

[16] Fourier, J. French mathematician. 1768–1830.

costs of the army and to increase the losses incurred by the enemy. About the same time, Dutch–American economist Koopmans (1910–1985)[17] formulated classical mathematical problems such as linear programming.

During 1946–1947, G. B. Dantzig[18] independently developed general linear programming and the *simplex method* for planning problems in the US Air Force.

The Problems

In the real world, there are many max–min problems involving linear inequalities. Let us consider the following problems,

Example 6.13 (Diet Problem):   There are m different types of foods $f_1, \ldots, f_m$, and there are n types of nutrients $n_1, \ldots, n_n$ that are needed for health. Let a_{ij} be the amount of nutrient n_j contained in one unit of food f_i. Let c_i be the daily requirement of nutrient n_i. Let b_i be the price per unit of food f_i. The problem is to supply the nutrient at the minimal cost.

We will set up a mathematical model as follows. Let y_i be the amount of food f_i used every day. Then we want to minimize the *object function*

$$b_1 y_1 + b_2 y_2 + \cdots + b_m y_m$$

subject to the following *constraints*

$$\begin{cases} a_{11} y_1 + a_{21} y_2 + \cdots + a_{m1} y_m \geq c_1 \\ a_{12} y_1 + a_{22} y_2 + \cdots + a_{m2} y_m \geq c_2 \\ \qquad \cdots \cdots \\ a_{1n} y_1 + a_{2n} y_2 + \cdots + a_{mn} y_m \geq c_n. \end{cases}$$

We certainly cannot purchase a negative amount of food, therefore we have

$$y_1 \geq 0, \quad y_2 \geq 0, \ldots, y_m \geq 0. \qquad \blacksquare$$

---

[17]Kantorovich, L. Soviet mathematician and economist. 1912–1986. Koopmans, T. C. Dutch–American mathematician and economist. 1910–1985. Kantorovich and Koopmans jointly won 1975 Nobel Prize.

[18]Dantzig, G. B. American mathematician. 1914–2005.

The above problem can be used in managing a chicken farm or an agriculture farm (foods become packed fertilizers).

Example 6.14 (Assignment Problem):   We have n workers and m jobs. Let the value of the ith person working on the jth job be a_{ij}. Let x_{ij} be the proportion of time the ith person spends non-negatively on the jth job. We want to maximize the following *object function* of total value:

$$\sum_{i=1,j=1}^{n,m} a_{ij}x_{ij}$$

under the *constraints*,

$$\sum_{j=1}^{m} x_{ij} \leq 1 \quad \forall i$$

with

$$x_{ij} \geq 0. \qquad\blacksquare$$

In 1947, when Dantzig showed John von Neumann his *simplex method*, von Neumann remarked from the point of view of *game theory* that there must be a *dual*. It turns out to be a fruitful remark. The dual problem becomes an integral part of the *primal* problem. We have the following definition of *standard problems and duality*,

Definition The dual problem of the *standard minimum problem*

$$minimize \ y^T b$$

subject to the constraints $y^T A \geq c^T$ and $y \geq 0$

is defined to be the *standard maximum problem*

$$maximize \ c^T x$$

subject to the constraints $Ax \leq b$ and $x \geq 0$.

Furthermore, the dual of the second *standard maximum problem* is the first *standard minimum problem*. $\qquad\blacksquare$

Certainly a minimum problem may involve $\leq$ or some y_j may not be non-negative. In those cases, we shall replace $\sum_i a_{ij}y_i \leq c_j$ by

$-\sum_i a_{ij} y_i \geq -c_j$ and $y_j - z_{1j} + z_{2j} \geq 0$ with $z_{1j} \geq 0$, $z_{2j} \geq 0$. Then the minimum problem is reduced to the standard form. Similarly we may reduce the general maximum problem to the standard form. Later on we show that in the interesting cases, the solution to a standard minimum problem is the same as the solution to its dual, and vice versa (cf. the duality theorem below).

We have the following definition to fix the terms we use in these problems.

Definition If only the empty set satisfies all constraints, then the problem is said to be *infeasible*. A point y is said to be a *feasible point*, if it satisfies all constraints. A problem is said to be a *feasible problem*, if there is at least one feasible point. A problem is said to be *unbounded feasible* if it is feasible, while the object function in the maximum problem have no maximum (or respectively, the object function in the minimum problem has no minimum). Otherwise it is called *bounded feasible*. ∎

We note that the intersection of convex sets is convex, and a linear inequality defines a convex set. Therefore the set defined by inequalities is a convex set. There are three possibilities:

(1) Empty set.
(2) The set is not empty and the object function is unbounded (for maximum problem it is ∞; for minimum problem it is $-\infty$).
(3) Bounded feasible.

A naive way of solving a bounded feasible linear programming problem is to find all vertices of the feasible set, and evaluate the object function at those vertices and pick up the vertices which yield the extreme values. This method works only for a small number of vertices. If n, m are sizable, then the number of vertices is very large. The number of the vertices is about $\frac{(n+m)!}{m!n!}$, here n is the number of inequalities, and m is the dimension of the space. If $n = m$ is large, then the number of vertices is $> 2^n$ which is exponentially large and beyond the power of a computer if $n \geq 100$. We study the whole problem and find an easy way, the *simplex method*, to solve a large-sized problem.

Let us work on the *standard minimum problem* to illustrate the method,

$$\text{minimize } y^T b$$

subject to the constraints $y^T A \geq c^T$ and $y \geq 0$.

We may introduce a set of *slack* x such that the problem can be rewritten as

$$\text{minimize } y^T b$$

subject to the constraints $x^T = y^T A - c^T$ and $x \geq 0$, $y \geq 0$.

The above equations can be rewritten analytically as,

$$\begin{cases} -x_1 + \sum_j a_{j1} y_j = c_1 \\ \\ -x_2 + \sum_j a_{j2} y_j = c_2 \\ \\ \vdots \quad \vdots \qquad \vdots \\ \\ -x_n + \sum_j a_{jn} y_j = c_n. \end{cases} \qquad (E)$$

Remark: Similarly we may have a *standard maximum problem* as follows,

$$\text{maximize } c^T x$$

subject to the constraints $y + Ax = b^T$ and $x \geq 0$, $y \geq 0$.

The above equations can be rewritten analytically as,

$$\begin{cases} y_1 + \sum_j a_{1j} x_j = b_1 \\ \\ y_2 + \sum_j a_{2j} x_j = b_2 \\ \\ \vdots \quad \vdots \qquad \vdots \\ \\ y_m + \sum_j a_{mj} x_j = b_m. \end{cases} \qquad (E')$$

■

The above system of equations (E) is essentially the reduced row echelon form of a system of equations. Recall that the variables y_j are called *free variables* and the variables x_i are called *fixed variables*, and in LP (linear programming) we call fixed variables as *basic variables* or *pivot variables* and the *free variables* are renamed to *non-basic variables*.

The separation of the *basic variables* and the *non-basic variables* are purely accidental. If $a_{ji} \neq 0$, we may rewrite the above equations by row operations to make the coefficients $a'_{ji} = -1$ and $a'_{jk} = 0$ for all $k \neq i$. In this way we switch x_i and y_j. We shall say y_j enters the basic variables and x_i leaves the basic variables. We shall name the above exchange as *switching x_i and y_j by the pivot* a_{ji}.

[Simplex Method]

Dantzig started the modern research on *linear programming* by inventing the *simplex method* in 1946–1947. He not only pointed out the *switching pivot* on a_{ji}, but also showed the goal of the switch and the process to achieve the goal. After many years of studying, mathematicians found that for some examples, the original process is not fast enough or even cycling to no end. The simplex method is modified to the *modified simplex method* (see below) to avoid the cycling phenomena.

Tableau

Instead of a system of equations we put all data in a board and call it a *tableau*. Let us work on the *standard minimum problem* to illustrate the method. We may form the following *tableau*,

	x_1	x_2	x_3	$\cdots$	$\cdots$	x_{n-1}	x_n	-1
y_1	a_{11}	a_{12}	a_{13}	$\cdots$	$\cdots$	$a_{1,n-1}$	$a_{1,n}$	b_1
y_2	a_{21}	a_{22}	a_{23}	$\cdots$	$\cdots$	$a_{2,n-1}$	$a_{2,n}$	b_2
$\vdots$	$\vdots$	$\vdots$	$\vdots$	$\vdots$	$\vdots$	$\vdots$	$\vdots$	$\vdots$
y_m	a_{m1}	a_{m2}	a_{m3}	$\cdots$	$\cdots$	$a_{m,n-1}$	$a_{m,n}$	b_m
1	$-c_1$	$-c_2$	$-c_3$	$\cdots$	$\cdots$	$-c_{n-1}$	$-c_n$	0

Remark: The *tableau* for the *standard maximum problem* is as follows,

	y_1	y_2	y_3	$\cdots$	$\cdots$	y_{m-1}	y_m	-1
x_1	a_{11}	a_{21}	a_{31}	$\cdots$	$\cdots$	$a_{m-1,1}$	$a_{m,1}$	c_1
x_2	a_{12}	a_{22}	a_{32}	$\cdots$	$\cdots$	$a_{m-1,2}$	$a_{m,2}$	c_2
$\vdots$	$\vdots$	$\vdots$	$\vdots$	$\vdots$	$\vdots$	$\vdots$	$\vdots$	$\vdots$
x_n	a_{1n}	a_{2n}	a_{3n}	$\cdots$	$\cdots$	$a_{m-1,n}$	$a_{m,n}$	c_n
1	$-b_1$	$-b_2$	$-b_3$	$\cdots$	$\cdots$	$-b_{m-1}$	$-b_m$	0

■

We illustrate the transformation of the above *tableau* for the *standard minimum problem* corresponding to the switch of variables x_i and y_j. First we will establish the aim of the *simplex method*. If $-c \geq 0$ and $b \geq 0$, then there are obvious solutions to the minimum problem as follows. We let $y = 0$, $x = y^T A - c \geq 0$, $b^T y = 0$, the negative of right-lowest corner value is the minimal value of the problem.

Firstly, we wish to transform the *tableau* to help us find a feasible point and then move from feasible point to feasible point to lower (at least not increase) the object functions.

Let us rewrite the switching in terms of the *tableau*. The legitimate way of transforming the above *tableau* is as follows.

We may interchange a variable y_j with a slack x_i if the coefficient $a_{ji} \neq 0$. Look at the ith equation of $x^T = y^T A - c^T$ which is

$$x_i = a_{j1}y_1 + a_{j2}y_2 + \cdots + a_{ji}y_i + \cdots + a_{jm}y_m - c_i. \qquad (1)$$

We switch the position of x_i and y_j and maintain the formality of the above equation as

$$(x_i' =)y_j = (a_{j1}/-a_{ji})y_1 + (a_{j2}/-a_{ji})y_2 + \cdots + (-1/-a_{ji})x_i$$
$$+ \cdots + (a_{jm}/-a_{ji})y_m - (c_i/-a_{ji}).$$

Let $y_j' = x_i$, $y_s' = y_s$, $a_{js}' = (a_{js}/-a_{ji})$ and $c_i' = (c_i/-a_{ji})$, then we have

$$x_i' = a_{j1}'y_1' + a_{j2}'y_2' + \cdots + a_{ji}'y_j' + \cdots + a_{jm}'y_m' - c_i'. \qquad (2)$$

For all other indices $k \neq j$, after eliminating y_j by substituting the above equation into the following equation

$$x_k = a_{k1}y_1 + a_{k2}y_2 + \cdots + a_{kj}y_j + \cdots + a_{km}y_m - c_k, \qquad (3)$$

we get a new equation

$$
\begin{aligned}
x_k &= (a_{1k} + a_{ik}a_{1j}/-a_{ij})y_1 + (a_{2k} + a_{ik}a_{2j}/-a_{ij})y_2 \\
&\quad + \cdots + (a_{ik}/a_{ij})x_j + \cdots + (a_{mk} + a_{ik}a_{mj}/-a_{ij})y_m \\
&\quad - (c_k + a_{ik}c_j/-a_{ij}) \\
&= a'_{1k}y'_1 + a'_{2k}y'_2 + \cdots + a'_{ik}y'_i + \cdots + a'_{mk}y'_m - c'_k.
\end{aligned} \qquad (4)
$$

The object function $f = \sum_i b_i y_i$ is changed to

$$
\begin{aligned}
f &= (b_1 + b_j a_{j1}/-a_{ji})y_1 + (b_2 + b_j a_{j2}/-a_{ji})y_2 + \cdots + (b_j/a_{ji})x_i \\
&\quad + \cdots + (b_m + b_j a_{jm}/-a_{ji})y_m + (b_j c_i/-a_{ji}) \\
&= b'_1 y'_1 + b'_2 y'_2 + \cdots + b'_m y'_m - v'.
\end{aligned} \qquad (5)
$$

We start work on the vector $-c$. We want to improve the value of it if necessary. We have case (1) $-c \geq 0$, and case (2) $-c_j < 0$ for some j. There are further two subcases for case (1): subcase (1,a): $b \geq 0$, subcase (1,b): there are some i such that $b_i < 0$. In the subcase (1,a), the minimum problem is solved by taking $y = 0$, then $x = y^T A - c = -c \geq 0$. Therefore it is a feasible point and $\sum b_j y_j - v \geq -v$ always for $y \geq 0$. We find the minimun which is $-v$. Thus if we move the *tableau* to reach subcase (1,a), then we find the minimal value of the problem and we are done.

Let us postpone the discussion of subcase (1,b) for a while. In case (2), let $-c_j < 0$ for some index j. We want to show that in the bounded feasible problem, there must be an index i such that $a_{ij} > 0$ (see the proposition below).

We have,

Proposition 6.5.1: *Suppose that $-c_j < 0$ and $a_{ij} \leq 0$ for all i, then the minimum problem is infeasible.*

Proof. The jth equation of $x^T = y^T A - c^T$ is

$$x_j = \sum_i y_i a_{ij} - c_j < 0$$

for $y \geq 0$; it contradicts the requirement $x_j \geq 0$. Therefore it is infeasible. ∎

We assume that the problem is bounded feasible and in case (2). Select $-c_j < 0$ (we usually select the most negative $-c_j$, the so-called *"most negative pivot rule"*), therefore we have an index i such that $a_{ij} > 0$. We use the index i, consider all those elements $a_{ik} < 0$ with $-c_k \geq 0$ on the ith row and compare $-c_k/a_{ik}$ with $-c_i/a_{ij}$, if there is any. Note that all comparing numbers are non-positive. We select one closest to 0, i.e., the largest one. Say we select a_{ik} (where k may be j), then we interchange x_k with y_i.

Let us examine the results of this interchanging of x_k with y_i. Let us discuss two subcases (2,a): $k = j$, and (2,b): $k \neq j$. In the first case, we have (a) $-c'_j = -(c_j/-a_{ij}) > 0$. and (b) for all $-c_s \geq 0$, we have $-c'_s = -(c_s + a_{is}c_j/-a_{ij}) \geq -c_s \geq 0$ if $a_{is} \geq 0$. Furthermore, if $a_{is} < 0$, then by our requirement, we prove that $-c'_s = -(c_s + a_{is}c_j/-a_{ij}) \geq 0$. We conclude that in subcase (2,a), it is getting better.

Let us consider subcase (2,b): $k \neq j$. We have (a) $-c'_k = -(c_k/-a_{ik}) \geq 0$ and (b) for all $-c_s \geq 0$, we have $-c'_s = -(c_s + a_{is}c_k/-a_{ik}) \geq -c_s$ if $a_{is} \geq 0$. Furthermore, if $a_{ik} < 0$, then by our requirement, we prove that $-c'_k = -(c_k + a_{ik}c_j/-a_{ij}) \geq 0$. (c) Finally we have $-c_j < 0$ and $-c'_j = -c_j - a_{ij}c_k/-a_{ik} \geq -c_j$.

In case (2), we conclude that the set $\{c_t\}$ is not getting worse, and may be getting better. The classic simplex method finds in all applications that it indeed gets better and finally we may assume $-c_j \geq 0$ for all j. It is clear that if $c_k = 0$, then $c'_k = c_k$, $c'_s = c_s$ and $c'_j = c_j$, there is no improvement on those critical constants. Note that when we set $y' = 0$, then we have $x'_k = 0$. Geometrically, there are at least $m + 1$ hyperplanes intersecting at the point. A change of m hyperplane among the set will not move the point, and will not change the values of the object function. In fact, mathematically there are *cycling examples*. Then Bland[19] came to the rescue with an article *New finite pivoting rules for the simplex method* (Math. of Oper. Res. 2, 1977, pp. 103–107): with the *smallest Subscript Rule*: if there is a choice of the pivot columns (or a choice of the pivot rows), select the column (or row) with the x-variable having the lowest subscript (or the y-variable having the lowest subscript). This selection of pivot avoids the *cycling*. This method will be named the *modified simplex method*.

With the above *modified simplex method*, we assume that $-c_i \geq 0$ for all i eventually. At this stage, the point with $y = 0$, and $x = A^T y - c \geq 0$ is a feasible point. Once we find a feasible point, we try to reduce the value of the object function if possible. We assume that we are in the subcase (1,b) now and work towards subcase (1,a).

We have the following proposition,

Proposition 6.5.2: *If we have $-c \geq 0$, $b_i < 0$, and $a_{ij} \geq 0$ for all j, then the problem is unbounded.*

Proof. Let us consider the points $y_k = 0$ for all $k \neq i$ and $y_i = L$ where L is arbitrarily large. They satisfy the equations $y^T A - c^{'^T} \geq 0$ and are feasible points. Apparently the object function $y^T b$ goes down to $-\infty$. ∎

We assume that the problem is bounded feasible and is in subcase (1,b). Select $b_i < 0$ (we usually select the most negative $b_i < 0$, the so-called *"most negative pivot rule"*), among the index j for a_{ij}, let us consider all indices k such that $a_{ik} < 0$. We shall further select an index k with $-c_k/a_{ik}$ closest to 0, i.e., the largest one. Say we select a_{ik} (where k may be j), then we interchange x_k with y_i.

Let us examine the results of this interchanging of x_k with y_i. Let us discuss two cases: (1) $k = j$, and (2) $k \neq j$. In the first case, we have (a) $b_i' = (b_i/a_{ik}) > 0$ and (b) $-c_r' = -(c_r + a_{ir}c_k/-a_{ik})$ which is ≥ 0 if a_{ir} is non-negative. On the other hand if $a_{ir} < 0$, then by our requirement, we prove that $c_r' = (c_r + a_{irj}c_k/-a_{ik}) \geq 0$. Furthermore,

---

[19]Bland, R. American mathematician and operations researcher. 1948–.

in case (2), we have the object function $f = \sum_j b_j y_j + v = \sum_j b'_j y'_j + v'$ where $v' = v + b_i c_k / -a_{ik} \le v$.

We conclude that the new point $y'_1 = 0$, $y'_2 = 0, \ldots, y'_n = 0$ stays inside the feasible set, and the object function $f = b^T y$ is not getting worse and likely to get better. Eventually we get the minimal value by the *modified simplex method*.

In practice the simplex method is $O(n)$ for most problems; there are some LP (linear programming) problems, for instance, the Klee–Minty examples, (cf. Exercise 9) that all solutions using simplex methods are slowing down exponentially. If we want to have a polynomial time method, we have to either modify the existing simplex methods or look for other possibilities. Before we do that, we shall write down the linear algebraic representations of linear programming.

[Linear Algebraic Representation of Linear Programming]

We shall reformulate the above *tableau* in terms of linear algebra. Let us consider that standard minimum problem. We want to minimize $y^T b$. We write the equation $x^T = y^T A - c^T$ as $-x^T I + y^T A = c^T$ or the transpose of it as

$$[-I | A^T] \begin{bmatrix} x \\ y \end{bmatrix} = c.$$

In the simplex method, we exchange the x-variable with the y-variable, which corresponds to exchanging one of the first m columns with one of the last n columns of coefficient matrix $M = [-I | A^T]$. Then we use a sequence of row operations to make the first $m \times m$ matrix $-I$ again.

Remark: Let us consider that standard maximum problem. We want to maximize $c^T x$. We write the equation $y = b - Ax$ as $yI + Ax = b$ or

$$[I | A] \begin{bmatrix} y \\ x \end{bmatrix} = b.$$

In the simplex method, we exchange the x-variable with the y-variable, which corresponds to exchanging one of the first n columns with one of the last m columns of coefficient matrix $M = [I|A]$. Then we use a sequence of row operations to make the first $n \times n$ matrix I again.   ∎

Let us study the standard minimum problem again. The sequence of interchanges x_i with y_j as represented by multiplication by a matrix E^{-1} from the left and we have $[-I/A^T]E = [-B/F]$. The row operations can be represented by a multiplication by a matrix B^{-1} from the left on $M = [-I|A^T]$,

$$B^{-1}([-I|A^T]E)\left(E^{-1}\begin{bmatrix} x \\ y \end{bmatrix}\right) = B^{-1}[-B/F]\begin{bmatrix} x' \\ y' \end{bmatrix} = B^{-1}c = c'.$$

Or

$$[-I|B^{-1}F]\begin{bmatrix} x' \\ y' \end{bmatrix} = B^{-1}c = c'.$$

To begin with, we may further add a last row to indicate the object function f as,

$$\begin{bmatrix} -I \mid A^T \\ 0 \mid b^T \end{bmatrix}\begin{bmatrix} x \\ y \end{bmatrix} = \begin{bmatrix} c \\ 0 \end{bmatrix}.$$

After the above operations, we have

$$\begin{bmatrix} -I \mid B^{-1}F \\ b_B^T \mid b_F^T \end{bmatrix}\begin{bmatrix} x' \\ y' \end{bmatrix} = \begin{bmatrix} c' \\ 0 \end{bmatrix}.$$

Now we shall reduce b_B^T in the last row to 0. Note that this is simply a substitution process.

The last row becomes $[0,\ldots,0|b_F^T + b_B^T B^{-1}F] = [0,\ldots,0|(b')^T]$ for the coefficient matrix on the left side of the equation, and on the right side of the equation, the last value is $[b_B^T B^{-1}c]$. Note that $(b')^T = b_F^T + b_B^T B^{-1}F$. The stopping conditions are $c' \geq 0$ and $b' \geq 0$. The minimal is reached when the program stops, we have that the minimal value of the object function is $-b_B^T B^{-1}c$.

[Interior Point Method]

The most prominent method other than the *simplex method* is the *interior point method*. Years ago, John von Neumann suggested an interior point method which is neither polynomial time method nor efficient method. Khachiyan[20] showed the *ellipsoid method* was a polynomial time algorithm (in fact $(n^6 L)$). In 1984, Karmarkar[21] introduced *Karmarkar's algorithm* to solve the LP (linear programming) problem. It requires $O(n^{3.5} L)$ operators on $O(L)$ digit numbers and is efficient in practice. There are now algorithms which have better computational complexity and better practical performance.

Let us introduce the momentous Karmarkar method. In principle it is similar to the *Conjugate Gradient Method* of the last section. Let us study the minimal problem. With the assumption that there is a feasible point, we may assume it is an interior point $e = [1, 1, \ldots, 1]^T$. Furthermore, we assume that the minimal value is 0. In general it suffices to know the minimal value of the object function. This assumption is difficult to fulfill. Then we find the direction of the steepest descending. We shall move along that direction and stop before reaching the boundary of the feasible set. This can be achieved by building a *barrier* along the boundary (the old *barrier problem*). A typical problem is as follows,

$$barrier\ problem\ P(\theta) \qquad minimize \quad c^T x - \theta \left(\sum \ln x_i + \sum \ln y_j \right)$$

where θ is a small positive parameter. A function as above will prevent x_i, y_j going to 0. Once we stop at a point P, then we use the *projective scaling* (which is non-linear) to transform back the whole problem with P goes to $e = [1, 1, \ldots, 1]^T$, and repeat the above process. We could show that the series of values of the object function approaches the optimal.

In the following-up part of this section, we will prove another important theorem in LP.

---

[20]Khachiyan, L. Soviet mathematician who worked and died in the US. 1952–2005.

[21]Karmarkar, N. Indian mathematician. 1957–.

[Duality Theorem]

Let us consider a standard minimum problem called $Primal(P)$ and its dual problem $Dual(D)$:

(1) Primal(P): Minimal $y^T b$, subject to $y \geq 0$, and $y^T A \geq c$.
(2) Dual(D): Maximal $c^T x$, subject to $x \geq 0$, and $Ax \leq b$.

The duality theorem is the following,

Theorem 6.5.3: *When both problems are feasible, then they have optimal y^* and x^*, and the minimal of the primal problem $(y^*)^T b = c^T x^*$ is the maximal of the dual problem.*

Before we prove the above theorem, we shall prove the following lemmas,

Lemma 6.5.4 (Weak Duality): *If y and x are feasible in the primal and dual problems, then $c^T x \leq y^T b$.*

Proof. We have

$$c^T x \leq y^T A x \leq y^T b$$

since $x \geq 0$ and $y \geq 0$. ∎

Lemma 6.5.5: *If y and x are feasible in the primal and dual problems, and $c^T x = y^T b$. Then both are optimal.*

Proof. Since for any feasible x', we always have $c^T x' \leq c^T x^* \leq y^T b = c^T x$, therefore $c^T x$ is the largest, i.e., optimal. Similarly $y^T b$ is optimal. ∎

Proof of the Theorem. Let us use the notations of the subsection on Linear Algebraic Representation of Linear Programming. Note that the simplex method always solve a bounded feasible standard min–max problem. Given a primal (i.e., a minimum) problem and having $-b_B^T B^{-1} c$ as the minimal value of the object function. Then we know that the vector $[x', y']^T$ satisfies the stopping conditions $b' \geq 0$, $c' \geq 0$. Especially we have $b' = b_F^T + b_B^T B^{-1} F \geq 0$ and the object function has the minimal value $= -b_B^T B^{-1} c$.

Let us select $\bar{x}^T = -b_B^T B^{-1}$. Then we have the value of the object function of the dual problem at the point $\bar{x}^T$ is $\bar{x}^T c = -b_B^T B^{-1} c =$ the minimal value of the primal problem. We only have to show that $\bar{x}^T$ is a feasible point for the dual problem, i.e., $\bar{x}^T$ satisfies the restriction for the dual problem,

$$x \geq 0, \quad Ax \leq b$$

or

$$x^T[-I|A] \leq [0,\ldots,0|b^T]$$

which is the same as making a permutation of columns by matrix E,

$$(x)^T[-B|F] \leq [b_B^T|b_F^T].$$

The first part follows from the definition of $\bar{x}^T = -b_B^T B^{-1}$, i.e., $\bar{x}^T(-B) = -b_B^T B^{-1}(-B) = b_B^T$, and the second part of the inequality follows from the stopping condition that $b' = b_F^T + b_B^T B^{-1} F \geq 0$, thus $b_F^T \geq -b_B^T B^{-1} F = \bar{x}^T F$. We conclude that $\bar{x}^T$ is a feasible point for the dual problem. Our theorem follows from the preceding lemma. ∎

Let us consider the following example,

Example 6.15: Let us consider a *primal problem*: some food for vitamin A are carrots (C) and sweet potatoes (S). Every pound of carrots (C) contains 4 units of vitamin A, while every pound of sweet potatoes contains 3 units of vitamin A, while every person needs at least 2 units of vitamin A daily. The prices of one pound of carrots and sweet potatoes are \$1 and \$0.94, respectively. We have, with $y_1 =$ the amount of carrots used daily, $y_2 =$ the amount of sweet potatoes used daily, then we have the following primal problem,

Primal(P): Minimal $y^T b = 1y_1 + 0.94y_2$, subject to $y_1, y_2 \geq 0$, and $y^T B = 4y_1 + 3y_2 \geq 2$.

We may convert the above problem to its dual problem

Dual(D): Maximal $c^T x = 2x$, subject to $x \geq 0$, and $Bx \leq b$, i.e., $4x \leq 1$, $3x \leq 0.94$.

We may interpret the dual problem as making bottles of vitamin A pills of one unit each, and x is the maximal price for it. It is easy to check that the minimal of the primal problem is $\$0.5 =$ the maximal price of the dual problem.

Exercises

(1) Consider the linear programming problem. Find y_1 and y_2 to minimize $y_1 + 2y_2$ subject to the constraints,

$$y_1 + y_2 \leq 2$$
$$2y_1 + y_2 \leq 5$$

and $y_1 \geq 0,\ y_2 \geq 0$.

(2) Put the following linear programming problem in the standard form. Find x_1, x_2, x_3 to maximize $x_1 + 2x_2 + x_3 + 6$ subject to the constraints,

$$x_1 + 2x_2 + x_3 \leq 10$$
$$20x_1 + 5x_2 \qquad \geq 23$$

and $x_1 \geq 0,\ x_2 \geq 0$.

(3) Find the dual to the following minimum problem. Find y_1, y_2, y_3 to minimize $2y_1 + y_2 + y_3$ subject to the constraints, $y_i \geq 0$ for all i, and

$$y_1 - y_2 + y_3 \geq 3$$
$$-y_1 + y_2 + y_3 \geq 1$$
$$2y_1 + 2y_2 + y_3 \geq 4.$$

(4) Show that for a maximal problem as follows,

$$maximize\ c^T x + v$$

subject to the constraints $Ax \leq b$ and $x \geq 0$.

If $b \geq 0$ and $c \leq 0$ (it is the same as $-c \geq 0$), then we stop. The maximal value is v and the point is $x = 0$.

(5) In the preceding problem (3), find the minimal for the primal problem and the maximal of the dual problem and show that they are equal.

(6) Prove the duality theorem starts with a maximum problem which is bounded and feasible.

(7) Maximize $5x_2 + 2x_3 + 4x_4$ subject to the constraints $x_1 \geq 0$, $x_2 \geq 0$, $x_4 \geq 0$ and with x_3 unconstraint, and

$$-x_1 + 5x_2 + 4x_3 + 5x_4 \leq 5$$
$$3x_2 \qquad\quad + x_4 \quad\; \leq 2$$
$$-x_1 \qquad + 2x_3 + 2x_4 \quad \leq 1.$$

(8) Maximize $3x_1 - 5x_2 + 2x_3 - 2x_4$ subject to all $x_i \geq 0$ and

$$x_1 - 2x_2 - 2x_3 + 2x_4 \leq 0$$
$$2x_1 - 3x_2 - 2x_3 + x_4 \leq 0$$
$$2x_2 \qquad\qquad\qquad \leq 1.$$

Show that after following the steps (1) $x_1 \Leftrightarrow y_1$, then (2) $x_2 \Leftrightarrow y_2$, then (3) $x_3 \Leftrightarrow x_1$, then (4) $x_4 \Leftrightarrow x_2$, then (5) $y_1 \Leftrightarrow x_3$, then (6) $y_2 \Leftrightarrow x_4$, then you are back where you started. This is *cycling*.

(9) (Klee–Minty example for $n = 3$) Maximize $100y_1 + 10y_2 + y_3$ subject to

$$x_1 \qquad + \quad y_1 \qquad\qquad\qquad = 1$$
$$x_2 \quad + 20y_1 + \quad y_2 \qquad\quad = 100$$
$$x_3 + 200y_1 + 20y_2 + y_3 = 10{,}000$$
$$y_1, y_2, y_3, x_1, x_2, x_3 \geq 0.$$

Show that the following using the "*most negative pivot rule*" for $-c_i$, it requires $2^3 - 1 = 7$ steps to reach the optimal value, (1) $x_1 \Leftrightarrow y_1$, then (2) $x_2 \Leftrightarrow y_2$, then (3) $x_1 \Leftrightarrow y_1$, then (4) $x_3 \Leftrightarrow y_3$, then (5) $y_1 \Leftrightarrow x_1$, then (6) $y_2 \Leftrightarrow x_2$, then (7) $x_1 \Leftrightarrow y_1$ then you reach the optimal.

(The original examples are: maximize $\sum_{j=1}^n 10^{n-j} x_j$, subject to

$$2\sum_{j=1}^{i-1} 10^{i-j} i x_j + x_i \leq 100^{i-1} \quad \text{for } 1 \leq i \leq n \text{ and } x_j \geq 0.$$

You need $2^{n-1} - 1$ steps to reach the optimal if you use the same pivot rule.)

Index

Printed in the United States
by Baker & Taylor Publisher Services